D0263191

LOOKING FOR ADVENTURE

30131 05038825 2
LONDON BOROUGH OF BARNET

Naturalist Steve Backshall is the presenter of the top-rated BBC natural history programme *Deadly 60*. After writing for Rough Guides and filming documentaries for the National Geographic Channel as their 'Adventurer in Residence', he joined the BBC, where he became a key member of the expedition team for the hit series *Lost Land of the Jaguar*, *Lost Land of the Volcano* and *Lost Land of the Tiger*. He has written three books for children. *Looking for Adventure* is his first adult title. He lives in Buckinghamshire.

LOOKING FOR ADVENTURE

STEVE BACKSHALL

PHOENIX

A PHOENIX PAPERBACK

First published in Great Britain in 2011
by Swordfish
This paperback edition published in 2012
by Phoenix,
an imprint of Orion Books Ltd,
Orion House, 5 Upper St Martin's Lane,
London WC2H 9EA

An Hachette UK company

1 3 5 7 9 10 8 6 4 2

Copyright © Steve Backshall

The right of Steve Backshall to be identified as the author of this work has been asserted by him in accordance with the Copyright, Designs and Patents Act 1988.

All rights reserved. No part of this publication may be reproduced, stored in a retrieval system, or transmitted, in any form or by any means, electronic, mechanical, photocopying, recording or otherwise, without the prior permission of the copyright owner.

Every effort has been made to fulfil requirements with regard to copyright material. The author and publisher will be glad to rectify any omissions at the earliest opportunity.

A CIP catalogue record for this book is available from the British Library.

ISBN 978-0-7538-2872-4

Printed and bound by CPI Group (UK) Ltd, Croydon, CR0 4YY

The Orion Publishing Group's policy is to use papers that are natural, renewable and recyclable products and made from wood grown in sustainable forests. The logging and manufacturing processes are expected to conform to the environmental regulations of the country of origin.

www.orionbooks.co.uk

To Mum and Dad,
for getting me started on this great adventure…

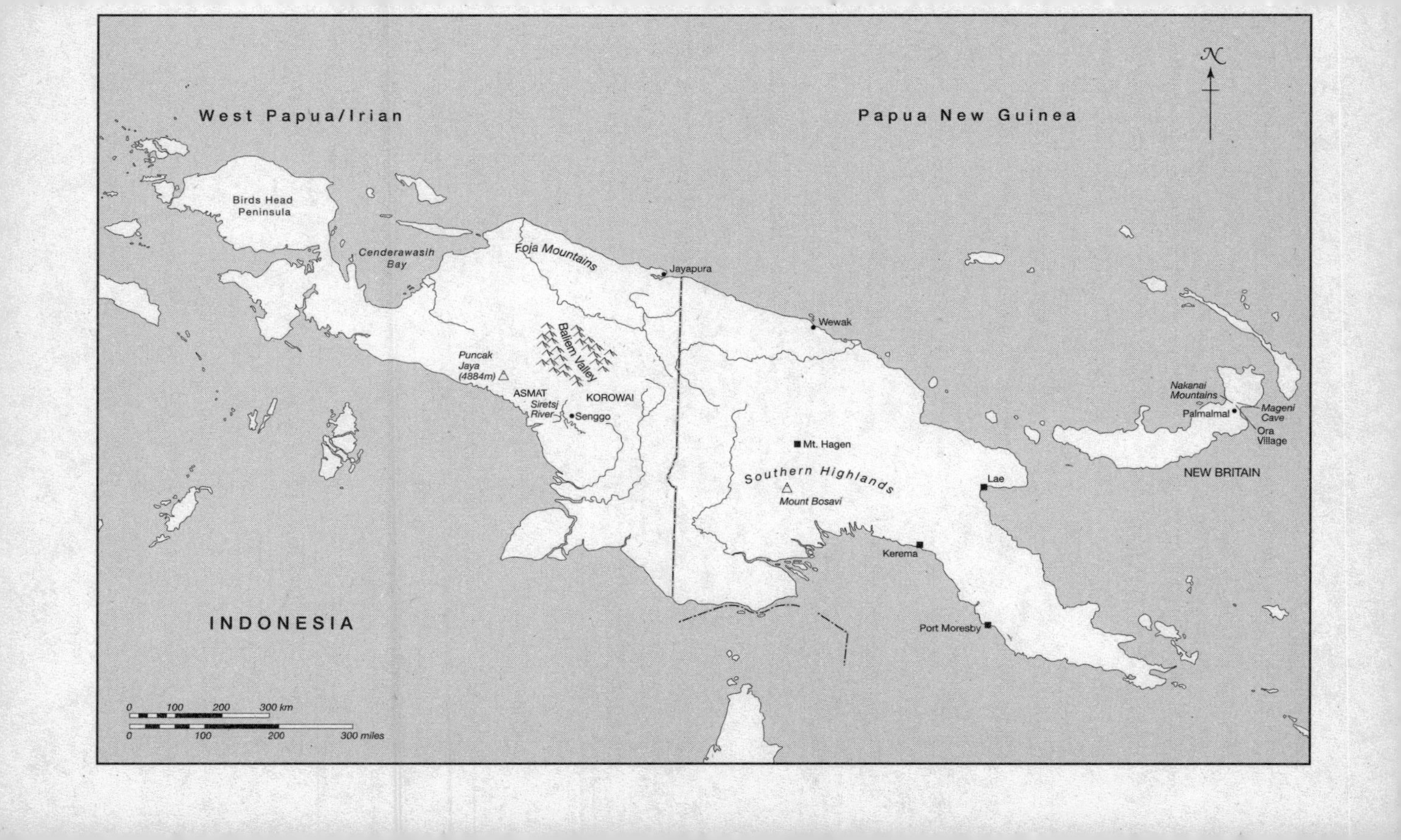
N
West Papua/Irian
Papua New Guinea
Birds Head Peninsula
Cenderawasih Bay
Foja Mountains
Jayapura
Wewak
Baliem Valley
Puncak Jaya (4884m)
ASMAT
KOROWAI
Siretsj River
Senggo
Mt. Hagen
Southern Highlands
Mount Bosavi
Lae
Kerema
Port Moresby
Nakanai Mountains
Palmalmal
Mageni Cave
Ora Village
NEW BRITAIN
INDONESIA
0 100 200 300 km
0 100 200 300 miles

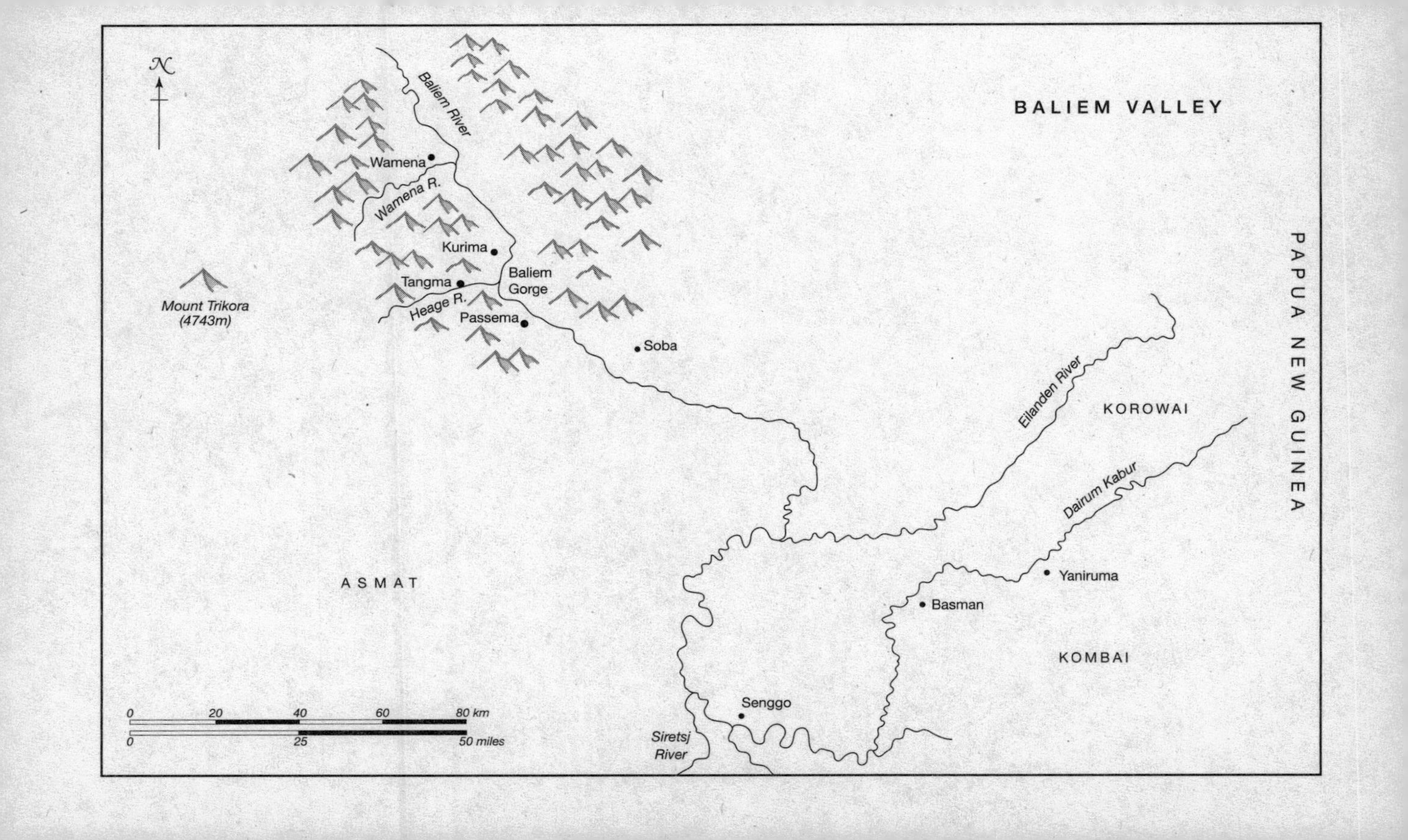
BALIEM VALLEY
N
Baliem River
Wamena
Wamena R.
Kurima
Tangma
Baliem
Gorge
Heage R.
Passema
Soba
Mount Trikora
(4743m)
ASMAT
PAPUA NEW GUINEA
Eilanden River
KOROWAI
Dairum Kabur
Yaniruma
Basman
KOMBAI
Senggo
Siretsj
River
0
20
40
60
80 km
0
25
50 miles

PART ONE

Prologue

'Get a load of that – that's her!' screamed our Aussie pilot above the roar of the rotors and the battering of the blessed cold wind on my face through the chopper's open door. The broccoli florets of the forest zipped just metres below the skids, as we flew steadily upwards over the mountain's perfect green lower slopes, before the steepness increased and all of a sudden the ridges of the outer caldera appeared, draped in a green crushed-velvet throw of forest. 'You must be out of your tiny little mind, mate! Tell you what, you and the jungle bunnies go on down in there with the leeches and the snakes – I'll be back at base cracking open a cold one.'

As we neared the rim of the volcano, the outer slopes became almost vertical. We swooped up to the edge, which appeared to be a finger's breadth wide – and then abruptly, nothing, a massive drop into the underworld. As the helicopter cruised over the rim and the abyss opened up beneath us, we all experienced a full-on roller-coaster moment, hearts dropping out the bottom of stomachs. It was so epic that every well-planned thing I was ready to say to the camera just went out of the window. 'OH MY GOD!' I blurted. Four kilometres across and more than a kilometre deep, Mount Bosavi's crater could have totally swallowed my home town.

From up high the inside of the cauldron looks like a perfect fruit bowl, but as you drop down inside it you become aware that it is strewn with imperfections. The inner caldera is in most places vertical rock, but there are some precipitous ridges running down to the bottom, and it's one of these we had hoped to descend to get us inside. The ridges were anything but smooth elevators, dropping down uniformly into the bowl: broken, tangled with forest, many came to abrupt ends in sheer drops and landslides. Right now, our grand plan seemed as well conceived as that of Burke and Wills, who back in the 1860s attempted to walk across the Australian outback with twenty tonnes of kit including a solid oak dining table, candelabras and zero bush experience.

Inside Bosavi's bowl there are also peaks and folds, many several hundred metres high, often with big drop-offs over rock walls. Getting round any one of those could take a whole dangerous day. Trees that I'd taken for spindly saplings from up in the chopper were actually fifty-metre tropical hardwood giants. Streams that appeared to be mere trickles from the height of the crater rim were now clearly rampaging hell rivers, with multiple drops over waterfalls and through boulder-choked gorges. My initial thought that we'd be able to follow a streamway was clearly hopeless optimism. There was nowhere instantly obvious inside Bosavi that you could land a helicopter, and if we did find a spot it would require several days of chopping and clearing to turn it into a landing site. Suddenly I began to see quite how badly wrong this apparently simple expedition could go.

The expedition was the focus of an international television series devised by the British Broadcasting Corporation to explore forgotten corners of Papua New Guinea, find new species and bring to the public eye the potential conservation nightmare unfolding in these hinterlands, far from the eyes of the international press. No outsider had ever ventured into Mount Basavi before; it represented a virtual island created by altitude, its near-vertical sides and its isolation held the promise of untold treasures and adventure beyond our wildest dreams. If there remained a Lost World on our shrinking and increasingly homogenised planet, this was it.

After about fifteen minutes of circling – the pilot with his eyes on the fuel gauge like a kestrel watching a vole – we had a breakthrough. Sometime during recent months a major landslide had plummeted down one of the smaller waterways, obviously propelled by a flash flood. The slide had taken with it tonnes of boulders and earth, as well as most of the vegetation that had covered the hillside. The resultant heap of bare mud and rocks looked to have consolidated, and was perhaps twenty metres across by fifteen. With a bit of clearing work, it was just possible that a very daring pilot would be able to land there. I suggested this to our pilot, who hovered over it for a few minutes, his face pursed in concentration. Eventually he nodded: 'You'd need to flatten it all out, and clear down all those trees, but then . . . it might work ... rather you than me getting down here though ... Hell, man!'

We hovered perhaps a hundred metres above it, as low as we could get with the forest cover, and I programmed the rough spot into my GPS as a waypoint. This was all we'd have as a guide to make our way down here. Our only landmark was a huge, branchless tree trunk so pale in colour it appeared totally white. Because the understorey had been cleared out by the landslide, it seemed like a ghostly spirit tree.

With those measly few coordinates logged, we felt our tummies drop again as the whirlybird lifted up and out of the crater, and back down the outer slopes of Bosavi. A substantial distance down one of the ridges was a clear spot in the trees, so our pilot dropped a skid down. With the rotors still whirring, we pulled our kit from the cab. Crouching beneath the blades, faces blasted by the windblown leaves, we dashed for cover. As we waited for the pilot to return with our local guides, Maxi and Chocol, a New Guinea harpy eagle flew overhead, letting out its extraordinary 'bonk bonk' call, like the plucking of a hunter's bowstring.

After the last chopper drop, we pressed on up the ridge line along a tiny overgrown path, the stunted trees wrapped in a duvet of dripping green moss that deadened sounds and poured pints of wet goo down your neck every time you stooped beneath a low-hanging branch. We reached the top as clouds began to roll in, minutes before the rains bombarded us, but in time to see the crater revealed in all its majesty. It was simply overpowering. If Spielberg had seen this place first, he would have filmed *Jurassic Park* here. The jagged, green crater rim was like an inside-out fortress. The mist and clouds raging up the crater walls made it seem as if the tree-tops were on fire, or as if the volcano itself was still smouldering. I was standing right up on the rim, and the world below me was a total unknown, unmapped, untrammelled, unexplored. I would very soon become the first white man to ever venture inside. The mist-infused jungle bowl below me was for now the undiscovered country from whose bourn no traveller returns. This truly was my enterprise of great pitch and moment!

It was a genuine surprise to find that, from nowhere, I was overcome by a sensation of overwhelming excitement and emotion. Suddenly the whole thing fell into place. For the first time I knew why I was here – it all had a purpose, a reason. I was here to make my peace with New Guinea, to lay to rest the ghosts of explorations past, and to rediscover and finally realise all my childhood dreams and fears. After earlier, humiliating, aborted expeditions here, I'd shelved New Guinea in a folder marked 'Suppressed'. Now, standing there on that ridge, I saw Bosavi as a great adventure long overdue, a chance to set the score straight with New Guinea and feel again the hunger for the place that I'd had for most of my life. With wondrous honeyeaters and fruit doves coursing about in front of me, ears alive to the cheepings and purrings of unseen frogs in the ferns and mosses, I stared down, mesmerised, into the crater below.

Seething tendrils of cloud billowed up the ravaged walls, as if witches were mixing some primeval potion in a giant green cauldron. Out beyond the mountain and off into the distance in every direction stretched a sea of cloud, with only a few taller peaks poking up through the perfect unbroken white blanket. Above the fierce whiteness, blue skies and a quickly rising sun. This is what is known as an inversion, where the usual thermocline – temperature gradient – is reversed, and cold air lies in the valleys with a layer of warm air above it. Water vapour condenses here, forming an ocean of bubbling cloud. Usually this will burn off after a few hours when the sun starts to redress the balance and air layers start to mix; but until it does, it feels like you are gods, sitting on the cumulus couches of your own private heavens above the clouded world.

Tiny colourful birds perch on the moss-covered boughs about us, chirping inquisitively at these strange intruders. Even my cynical godless Western mind can see why this place holds such powerful magic for the local people. To further emphasise the point, Chief Seguro of Sienna Falls sings us into the crater, creating a gate using a magic stick. He chants a refrain to the gods of Bosavi crater, to a giant serpent said to dwell around the rim and to the tree kangaroos within, imploring them to give our journey their blessing. I can feel the thumping in the back of my throat as we cross under the gateway, beneath our feet the unremitting steepness and miracles unknown.

The heart of Bosavi steamed and bubbled. Of course the crater is long extinct, and the steams boiling up from below were merely mountain mists enveloping the trees within. The mountain itself stood above the clouds, themselves a bleached white carpet stretching out towards the horizon in every direction. Beneath the cloud, the forests breathed and a billion creatures known and unknown stirred with the sunrise. But below me was the prize. A forgotten world shut off from the outside world through time, unexplored by scientists or travellers, unknown to all but the tribes that live here. My next steps would carry me down into the crater, into the Lost World inside the volcano.

1

The jungle-veiled mountains are a span of infinite, mist-saturated wilderness outside my smoky plane window; dark-brown rivers carve through the valleys, reflecting the sun in sudden unexpected flashes of blinding gold. We had been flying over this landscape for an hour, and at this height the lush forest appeared like a green, velvet rug over the studded peaks. Occasional flowering vegetation spangles the canopy below with bursts of yellow and white. From up here it looks as if you could surf over the emerald waves that are the mountains, but I know that the terrain below is as tough as jungle gets – to hike even a mile up the tangled slopes would be a hard-fought achievement. Within these forests, though, were hidden all the animals I had long dreamed of finding, along with muddy, rotting, mosquito-blown torments beyond imagining. American aeronautical maps of the area delineate the land below without contours or features – merely as blank white areas the size of Greater London and bearing the legend 'Relief Data Incomplete'. This tantalising unknown is not barren wasteland, but pristine jungle and tangled alluvial mangrove swamps that have never borne the imprint of an explorer's boot.

My name is Steve Backshall, and for the last fifteen or so years I have been travelling the world's most exotic and least-known places for a living – which sounds to many people like paradise. Yet at some level I am, and have always been, deeply unsatisfied. Nagging at the back of my mind has always been the sense that I'm searching for something but have never really put my finger on what that something might be – well, until I sat down to write this book. It had been on my mind for the best part of a decade while my old faux-leather-bound travel diaries gathered dust in the loft, when I finally managed to catch the interest of a West London literary agent.

In his plush office he lounged back on burnished leather, bookshelves stuffed with the writings of his eminent clientele. Adjusting his designer specs on his nose and sweeping back his fashionably long hair, he asked me a simple question: 'Why? Why do you do this with your life? What makes you want to continually push yourself? Most people are going to think you're completely insane.'

That was an easy one! I opened my mouth, confidently prepared to give him one of my many stored-up tales of derring-do, but faltered. Nothing came out. I was totally stumped. In the previous hour I'd already talked about how I'd been influenced by the great naturalists Gerald Durrell, Alfred Russel Wallace and David Attenborough, then about the cheap silver-fronted paperbacks of Wilbur Smith and Jack London, but to this important question I had no answer. Why is it that everything in my life has had to be a challenge? Just what is it I've been striving for? What's been the point of it all? I sat there like a guppy, mouth opening and shutting and no doubt with a stupid look on my face. I fully expected the answer to come to me on the train home, but it didn't. In fact, it took me nearly six months to figure out what it is that's been driving me all this time and put it into words, and now I have it in black and white it all seems a bit silly. But here it is.

I wasn't supposed to be here. I've been dumped out of time. I'm an anachronism, a refugee of circumstance trying to find my place in a shallow, shrink-wrapped modern world.

I know I'm far from alone. Practically every day that passes sees another of my competitors or colleagues paraglide off Everest in a gorilla suit or sail across the Pacific in a bathtub in a desperate search to affirm themselves, and to be the first at something. Anything! A couple of hundred years ago we would perhaps have been real explorers, discovering a world awash with undreamt-of miracles. However, modern explorers are not burly machete-wielding adventurers but big-brained scientists in search of 'the God particle', cold fusion or the cure for cancer. In fact just about everything about the modern world and its silicon chip, silicon chest, fast food, concrete and shining glass exterior leaves me cold, and with a deeply uncomfortable sensation in my belly. Could it be that I'm a savage dumped in modern times like Crocodile Dundee – well, except without the wit, the giant knife or the super-fit American girlfriend? Could this be why I have the niggling sense in the back of my mind that everything around me is somehow wrong?

Walking the Saturday streets of London, assaulted by the Chelsea chiquitas with their uniform big hair and surgically attached mobile phones, as white-van-man over-revs at the traffic lights and the newsstand headlines scream out about the antics of some rapist billionaire footballer as if it really matters, with technology all about me that would have been condemned as works of witchcraft not so long ago, I'm as lost as most people would feel if they were teleported to the middle of the Arctic tundra. Something went wrong with the space – time continuum and dumped me here at the dull end of the twenty-first century, struggling to find meaning in a world that seems to have none.

I always wanted to have been born in the Golden Age of Exploration, the mid-1800s, when Darwin and his contemporaries were sailing off into the unknown, never knowing if the Kraken would awake from the depths and eat the cabin boy. Some of their shipmates must still have believed they might sail off the edge of the world, and the real dangers of storms, starvation and cannibals' cooking-pots gave every day an edge you just can't replicate today – not unless you fancy wandering through Glasgow dressed as an American tourist and wearing a fat fanny pack. The great wanderers of yesteryear drew their maps as they went, discovering new wonders with every divine new day.

That was the time for me.

But instead, I was born in Bagshot in Surrey, the heartland of normal in the south of England, and lived my formative years through the eighties, the emotionally bankrupt decade that style forgot. The options open to me (though I've never rationalised it till now) were to sit around fantasising about going back in time, or to try and make a life less ordinary, to try and be an explorer now – in as much as that is possible in our modern world. This has been both my blessing and my curse in life. It's taken me to over a hundred countries, driven me to climb to the tops of the highest Himalayan peaks, got me shot at, arrested and robbed blind, had me more frightened and more ecstatic than I could ever explain to anyone. However, it also leaves me here in my mid-thirties, single, restless, unsatisfied, sharing precious little common ground with my contemporaries and now contemplating where on earth I go from here, what on earth the point of it all has been, and most of all where it all ends. In the words of an astute friend: 'You're an island Stevie B.'

Even as a tiny child this dominating facet of my character was evident to all who knew me. My parents describe me as an India-rubber handful, who'd bounce out of his cot at every opportunity to try and eat my way through their best glassware, attempt to cut out my sleeping sister's tonsils with the best scissors, or climb the crockery cabinet in search of the cookie jar. I pulled the whole thing off the wall in an avalanche of ceramics – my parents were so happy to find me alive in the debris I escaped punishment! Far from trying to quash my adventurous spirit, Mum and Dad saw it as their mission to give my sister and me a childhood of extraordinary experiences, filled with challenges, triumphs – and traumas.

As I'm sure is true of most kids, my most powerful memories are from family holidays. While most of my friends would be taken away to a Butlin's holiday camp or the south of Spain, we were fed on a holiday diet of frugal adventure, which more than once meant nearly floating away on an airbed inside a drowning tent. Weirdly, it was on one of these camping trips that I first became aware of the big, scary dark place that was to have such an impact on my life. In my memory, we were camping in the New Forest, but in fact it could have been in the Welsh valleys or in a farmer's field in Cornwall; we rarely holidayed in the same place twice. It must have been raining really hard for several days (Dad would never have let us be inside, otherwise), to force us into a village hall with an exhibition on. The exhibition was all about Papua New Guinea.

I can picture my sis whingeing her way around the glass cases; all she ever really wanted to do was be with horses and it would have been her idea of hell. None of my family even remember the exhibition now, but it engraved itself in my imagination like a William Blake etching. The contents of the cases are a photo album of images that haunted my dreams and nightmares for the rest of my childhood. There were sepia-tinted black and white photographs of tiny muscular Papuan pygmies who came up to the waists of the white explorers who stood amongst them. The tribesmen really did have bones through their noses, their heads were crowned with brightly painted feathers and their groins sprouted penis gourds. Next to them were old-fashioned drawings of birds of paradise, their avian heads dipped in malachite and chartreuse oil paints, wearing Elizabethan ruffs round their necks and trailing extravagant wires and bustles. Surely they couldn't be real? There was a photo of a giant crocodile shot dead, its jaws held open by a white man with a rifle. Its head alone was over a metre long and it could have swallowed me in a single gulp – indeed, it would have had to, crocodilians being physically unable to chew.

The next case down held a row of shrunken heads. Shrinking heads is a deeply macabre process. Having viciously killed the owner, the headhunter cuts the head open at the rear and removes the skull. He then sews the eyes and lips shut with palm threads, before boiling the head in tannins and cooking the now leathery skin in hot sand. The effect is deeply disturbing, as the head is still unmistakably human. The skin is tanned dark as a well-worn Chesterfield, the nose and other features are huge and distorted, and the hair is still long and luxuriant a hundred years on. But now the whole head is only the size of a grapefruit. (This bit might be my memory messing with me, as the process of shrinking heads is more common in the Amazon basin than in Melanesia, but I still see them grinning back at me in my memory's blurred etching.)

The last glass case was the one that really held my attention, and if I shut my eyes I can still envisage it most clearly now. It was full of pinned insects from the Papuan forests. Until then, bugs to me were just – well, bugs. Ugly yucky things that ruined your picnics and seemed designed purely to make life a bit more uncomfortable. This lot, though, were a whole different world of horror. There were amethyst, amber and garnet-inlaid beetles, yellow- and black-striped spiders the size of my dad's mighty, meaty hand, and armoured scorpions whose sting must surely have meant instant death.

More scary than the scorpions, though, were the biggest, most monstrous insects I had ever seen: a foot long, and taking up half the case with their huge bulk. I know today that they were harmless stick insects, pinned with their spindly legs and colourful wings outspread to give them maximum effect, but at the time they were the most horrific things I had ever seen. Following the visit to the exhibition I began to wake up at night, tormented by a nightmare that had me lying in the jungle leaf litter, wrapped up in a sleeping bag, with those vast stick insects crawling over my face looking for a place to begin feeding.

After the Dark Island had first been installed in my consciousness, throughout my childhood every tiny brush with New Guinea cemented my fascination with the place. My mum brought back piles of *National Geographic* magazines from car boot sales, and I'd thumb threadbare the pages about the Star Mountains, the Sepik River and singsings in the misty highland valleys. My treasured encyclopaedia of birds of the world I still have – now in pieces, its spine broken through overuse in such a way that it opens automatically at the pages featuring the birds of paradise. I learnt everything about them, from woefully inadequate adjectives describing them – 'the *superb* bird of paradise', 'the *splendid* astrapia', 'the *magnificent* riflebird' – to their courtship dances, and pictured myself as the young Attenborough sitting beneath their *lek* (dancing site), as the birds danced above me spotlighted in a sunray they'd created for themselves by plucking a hole in the canopy above. I couldn't have told you where Manchester or Leeds was on a map of the UK, but could have directed you straight to the obscure Bomberai Peninsula, the Aru Islands or the spot where Jan Carstensz in 1623 first reported glimpsing an equatorial glacier (for which he was roundly ridiculed by all back in Europe).

At eighteen years old, having finished my A levels and saved up some money from making kebabs, tedious nightclub and market-stall jobs, I set off to Asia on a voyage of discovery.

For the first couple of months I fell with other backpackers into the easy, hedonistic comfort blanket of Thailand, which offered thousands of other lost and afraid young people the chance to travel without a scrap of effort or genuine risk. To begin with I was massively insecure, and as keen as anyone else to stick to the tourist trail. I drank too much, partied lots, and kidded myself that I was finding myself. However, as my confidence grew I realised that I wasn't actually achieving anything of worth, and more importantly that the travellers around me that appeared the coolest were no different from the cool kids back at my school. For much of my childhood I'd struggled to fall in with the cool set, but never really managed it. Now suddenly I saw them all with fresh eyes – subject to a lifelong bluff, a fixation with image concealing zero real substance. In a minor epiphany, I also realised that I preferred my own company to them, and that they were no more than a millstone round my neck. It's been a continuous battle against true misanthropy ever since.

For the next six months, though, I didn't fight it. Instead, I undertook my travelling like some frantic school project, taking myself on my own to the furthest-flung places imaginable, forcing myself to do it the hardest way possible, while forbidding myself luxuries like some sadomasochist monk. Reading back through my diaries now raises the hackles on the back of my neck – how I didn't get myself into more serious trouble God alone knows. I'd just hitch-hike out into the middle of nowhere, then start walking, sleeping wherever the day ended. Despite being robbed at gunpoint twice and stranded in nowheresville backwaters for weeks on end, I somehow managed to end up in many villages that had never seen white people before, and was the guest of honour at festivals and funerals of spine-tingling primordiality. It's a miracle I came home at all, really, but my university place was booked, and despite all efforts to scrimp and save, my finances eventually ran dry.

After finishing my three years at university, I looked around in dismay as my friends were sucked into jobs in management consultancy and insurance. They seemed to get middle-aged almost overnight. This was certainly not part of *my* life plan. My instant reaction was to do something as tangential as possible, so I took myself off to Japan. There I spent a year studying martial arts, teaching confused snotty-nosed Korean kids (who were still learning Japanese) to say 'apple' in English, and working as a model and extra in various television commercials.

It was a bizarre year, almost worth a book in itself, though spectacularly irrelevant to the things that I really wanted to do. I returned to the UK as a black-belt, convinced that my mission was to set up a dojo in Britain and get on to the British judo team. This grand plan lasted about six months, and was well countered by the fact that my parents were rather despairing over the fact that their son was seriously considering dedicating his life to – well, to beating people up. I suppose the fact that I never made the British judo team, or anything like it, probably had a bearing on things too. So it was with considerable excitement that over breakfast one day my mother showed me a little ad in the *Guardian* newspaper asking for writers to compile the first *Rough Guide to Indonesia*, a country I'd already had an introduction to on my earlier travels.

It seemed at that moment to be the gateway to a future of incredible possibilities. Leaving my scrambled eggs to congeal on the plate, I started compiling a CV and sample piece right there. There were several thousand applicants, but I doubt any of them attacked the preparation as I did. For a week I hid away in my parents' garden shed with their hundred-year-old word processor (another car boot sale special) preparing my sample chapter, even tracing the exact font the *Rough Guides* used and relearning the decent spoken Indonesian I'd learnt in the four months travelling there during my big year out. Talk about using a sledgehammer to smash a walnut! They'd never before seen a job applicant so over-prepared, and my naive keenness did me no favours. In the interview I told them I would do the job for free.

Little did I know that's pretty much exactly what I'd have to do, guidebook writers ranking some way behind street cleaners when it comes to cash.

Indonesia the nation had great appeal for me, but there was one area in particular that had me beside myself with excitement. During all my earlier travels, West Papua, then known as Irian Jaya, had appeared like the Holy Grail: the most out-there place on the continent, a final frontier in an ever-shrinking world. Unfortunately back then it had been way beyond my feeble budget. Now that my flights were paid for, though, I was going to make the absolute most of the opportunity, and I set aside three months to travel in Irian while putting together my chapter. At the time, the chapter was very much secondary – I was in my early twenties, and was only really interested in the allure of adventure and exploration that Irian offered. This was the place where the young TinTinesque Backshall could pit himself against the environment and win, gaining enough kudos to genuinely be able to call himself an explorer. Not only that, but the vast jungles held biological riches beyond my wildest dreams. This was not just one of the world's great hotspots of biodiversity, containing millions of species of exotic invertebrates and bizarre mammals, but was one of the most important places on earth for the history of the biological sciences.

It was the great zoologist Alfred Russel Wallace (1823–1913) who was the first to notice that as you travel east through Indonesia the flora and fauna change dramatically, with a big change happening about the middle of Indonesia. The basic premise is that New Guinea once shared a land bridge with Australia – maybe as recently as the last ice age – while Java, Sumatra, Borneo and Bali were connected to Asia. The region where the crossover occurs has been dubbed Wallacea, in Wallace's honour. Even today the gulf between the two small and comparatively close islands of Bali and Lombok is extraordinarily deep, and travelling between them you feel as if you are leaping from one continent to another. The vestigial wildlife is a result of these two ancient connections: from distinctly Asian in the west with large mammals such as tigers, rhinos and elephants, to distinctly Australasian in the east with marsupials and monotremes (mammals that lay eggs).

At this point I have a confession to make. I'd not even heard of Wallace until I started researching for the *Rough Guide*. His book *The Malay Archipelago* kept cropping up in my research, so after putting it off for some months I took it away on a trip, knowing that seclusion with no other reading material would force me to slog through it. In the event, I read it through as breathlessly as if it had been a thriller, then went straight back to page one and started again. It remains the single most inspiring book I have ever read. Though you couldn't dispute the genius of Darwin, his journals are all rather dry, with little flavour of the places he visited and little sense of the adventures he must have experienced. Wallace's books, on the other hand, have more swashbuckling and derring-do than Rider Haggard and Arthur Conan Doyle's back catalogues put together. This was a man who in the mid-nineteenth century put his life in the hands of headhunters and cannibals, discovered new and wondrous species of animal every day of his travels, and hacked his way through terrain that even today is some of the most thankless on the planet. His collections from the trip totalled more than 125,000 specimens and thousands of new species.

New Guinea was to Wallace what the Galapagos Islands were to Darwin: a place where the greater picture of how the world's species had developed became increasingly evident, and the grand idea of evolution became a – to many – uncomfortable but unavoidable truth.

Wallace spent most of the 1850s and 1860s travelling through eastern Indonesia, and lived for several years with the natives in New Guinea. After all those years studying the wildlife of the region, while in a malarial fever in the Spice Islands he was suddenly struck with an epiphany: the theory of evolution through natural selection. The older and more cautious Darwin was probably several years from publishing his theory when Wallace sent him his own thesis, hoping for advice. The fear of being gazumped was apparently what spurred Darwin into publishing his first papers on the origin of species, and bringing perhaps the greatest idea of all time into the public domain. This aspect alone of Wallace, his astonishing but overshadowed effect on science, is enough to make him my fondest hero. However, it was his travel writings that cemented my fascination with New Guinea. I would give anything to have been with him on those journeys. But as this was unlikely to happen outside of my dreams, my only option was to follow in his footsteps.

Planning the trip, however, was to prove more challenging than any I'd undertaken so far. This was long before the days of Google and email, so I took the train up to London every day and sat for weeks on end in the British Library and the School of Oriental and African Studies library, poring over texts on the area. My best sources were the decades-old texts of Michael Clark Rockefeller, and manky dust-mite-ridden tomes on primitive art and anthropology. It made me feel scholarly and serious, but was largely pointless: practical information on actually travelling out there was nonexistent.

On the map, New Guinea resembles a gigantic turkey, with a wattled throat and blunted beak. The eastern half of the island is the independent country of Papua New Guinea, while the western half is what was known until 2007 as Irian Jaya (now West Papua), under the control of Indonesia. What really strikes you when you look at a map of New Guinea, though, is quite how little is on it. A map of the UK shows towns, cities, roads, motorways and railways. A map of New Guinea is just green.

If you get a good map, you may find different shades of green – some to differentiate between the green of the lowland swamps and the mountainous forests of the central spine of the island – but there are barely any signs of human intervention. There are no more than a few roads around Irian's capital of Jayapura, which, if we stick to the idea that New Guinea looks like a giant turkey, lies about where its upper back would be. It's on the north shore, a fourteen-hour journey from Indonesia's capital Jakarta – fourteen hours in a plane, and you're still in the same country! Apart from those few roads, and a couple of tracks snaking out from two or three other small towns, there's nothing. Small villages are marked as tiny circles on the map, but those circles are stranded off in the mountains and swamps, often a hundred miles away from their nearest marked neighbours. If you want to get there, you walk, you paddle a dugout canoe, or you rely on the few airfields cut roughly out of the forest.

If I was going to rediscover the halcyon days of exploration anywhere in the modern world, surely this was the place.

After my fourteen-hour flight from Jakarta, my feet first touched Papuan soil in the Irian capital city. Jayapura itself will never be considered alongside Paris and London in the pantheon of great capitals. It lies at the bottom of partially forested hills, in the gully carved out by its own filthy river. You'd think the tangled mess of rusty-roofed shacks and crumbling mosques had been swept down the river gully by a huge flood, then dumped there unceremoniously just short of a merciful burial at sea. Despite the wispy sea breezes, the town is sticky sweet-hot, and the litter and human filth in the streets give off a putrid hum.

Yes, it's a bit grim, but I've trodden streets like this throughout Southeast Asia over the years, and if anything I feel more at home in places like Jayapura than in any Western towns or villages. The rules are pretty simple, and despite the grottiness and poverty, people are generally smiley and friendly and will welcome you into their homes like old pals. However, though Jayapura on the face of it seems very similar to all those other places, as I walked the streets – sketching a map of it as I went and sticking my nose into several less than salubrious eateries – there was no mistaking an uneasy undercurrent to the town. To begin with, I thought I was just being a bit of a big blouse, alone in a new place and feeling vulnerable. But the more familiar I became with it, the more certain I became that not just Jayapura, but Papua as a whole, had a deeply sinister atmosphere about it that left me feeling unsettled.

The country itself is unremittingly hard, and will always live in my memory as a cavern-like jungle drenched in rain and mud, a place where the sun is always a stranger; but it's more than that. Perhaps it's the fact that the people look so bloody scary. Whereas most Asian peoples are slender – almost fragile-looking – Papuans look as if they could pull your arms from your sockets and beat you to death with them. Though short in stature, their Melanesian physiques are powerfully muscular and lean. Flat-nosed and heavy-browed, they usually wear a serious expression, not to say a menacing glower. Even the old women look like they could step into the ring with Mike Tyson and give him a run for his money. And while most Asian peoples are deferential, studiously avoid conflict and fit a vaguely colonial model of etiquette, Papuans stare at you with often fierce and hostile eyes, unashamedly studying you and everything you're doing, distrustingly, and sometimes from mere feet away.

Often, in moments of social tension, Papuan people will take to brandishing their gigantic machetes, axes and spears, wantonly chopping at tree stumps and house timbers. It's thoroughly intimidating. Here in the cities, though, I think the unease I feel is mostly down to a genuine unfairness and inequality between locals and migrants. As I've said, native Irianese people are distinctively Melanesian in appearance, with tight frizzy hair and handsome flat-nosed features. However, all the people you see running the shops, wearing uniforms and riding the buses are slender and tall, with glossy straight hair and Indian complexions, and very obviously come from further east in Indonesia.

Continuing my wander through the streets of Jayapura, I'd frequently come across the sad sight of an old Papuan man or woman sat there naked, skin ravaged by various tropical diseases, with a few ropey-looking vegetables on the pavement in front of them for sale. I never saw anyone even stop to look at their wares, let alone buy any, but I did see several armed policemen forcibly moving one elderly man along, though he wasn't doing any noticeable harm. He got up from the muddy gutter with the difficulty of old age, and the silent acceptance of one who's become used to such indignity. Rather than going to his aid, I scribbled the story in my omnipresent notebook, then returned to my sleazy hotel, passing the half-hearted attentions of the drunken Filipino hookers who sprawled, sweaty and sexless, over the lobby sofas in their tatty nightgowns.

My windowless corrugated-iron room, with its pathetic sputtering fan and ancient snooker-table green baize carpet, would at any other time have been a hell I'd have given anything to avoid. However, I'd written everything that could ever be written about Jayapura, reviewed every hotel and whorehouse, had my stomach turned in every *warung* (restaurant), got the lowdown on all the buses, boats and planes leaving the city, and visited the tourist attractions. These consisted of a crocodile farm no one could direct me to, and a statue of General MacArthur looking down on a lake that supposedly held six-metre-long freshwater sawfish. With my work done and nothing outside to attract me, I sprawled on my thin bed for a siesta (which would certainly not help my jet lag), staring at the cockroaches on the ceiling and wondering what the hell had made me think Irian would be a good idea.

One of the first jobs I had to do on arrival in Jayapura was head to the district police station, registering my presence as a foreigner. The only way I would be allowed to progress further inland was to buy a *surat jalan* – literally, 'walking papers'. On these papers I had to log every destination I was hoping to visit, and upon arriving there I'd have to check in with the local police. It would be down to their discretion as to whether I would then have to endure (at my own expense) an armed Indonesian military guard. The head policeman wore black knee-high leather boots snug round his calves, tight khaki-green trousers, shirt tucked into his belt causing a sizeable pot belly to spill out. It was as if he'd been squeezed into his boots and trousers like a long balloon clenched in a fist. Clearly, he had nothing better to do with his day than keep me sweating.

'Why you want go here?' he demanded, pointing to my request for the southern jungles of the Asmat region.

I tentatively explained that the jungles of the south were supposed to be very beautiful.

'There nothing there, just dirty *hutan* [forest].'

'Very beautiful,' I said, 'birds of paradise, cuscus [a small marsupial], big crocodiles ... I like animals.' He snorted with incredulous laughter, drew on a *kretek* (a clove cigarette) and blew the fragrant smoke skywards. As he twisted the cigarette in his fingers, he ostentatiously revealed the nails of his little fingers, as long and manicured as a Singapore madam's. This is a typical affectation – it proves you're a wealthy man of leisure who doesn't ever have to do manual work. It also gives the wearer a handy tool for excavating an earhole or nostril.

'You don't want to go here,' he tells me. 'You want to go to Biak Island. Beaches, *scuba diving*, pretty girls. Many white people. They have animals there in a zoo. You should go to Biak Island.'

'I am sure Biak is very beautiful, but I came here to visit the Asmat region. I've travelled many days to get here from my country.'

'The people there are animals,' he replied matter-of-factly. 'No car, no clothes, no lights, no house – people eating people. Animals.'

In the back of my head, I could hear a voice demanding that I tell this arrogant scumbag what I thought of him and his prejudices – the same voice that nags me incessantly to scream out something obscene at the theatre during particularly quiet and emotional scenes. However, it was obvious that all I could do with this man was pamper his ego, be as subservient as possible and hope he got bored of playing the big boss with me. There's no doubting, as well, that his perception of Irian and its wonders is the common viewpoint amongst Java-centric Indonesian people. They see themselves as part of a modern international country, and consider it as a terrible embarrassment that in far-off corners of their nation people still live such backward lives.

'Here no good,' he said, crossing out an intended destination close to the border with Papua, and 'Here cannot', slicing my expectations with a sweep of his biro. 'Why want to go here? You must go to Biak Island, very nice place, white people like beaches.' And pointing at me firmly: 'You go to Biak Island.'

I felt my hackles rising. If there's one thing that I cannot abide, it's people telling me what to do. I know I'm not alone in this. I would cross continents to do the exact opposite of what someone had just told me to do, sometimes in a way that is just plain bloody-minded and may even negate common sense – but hey, it's got me where I am today and I'm not going to change now! Many people, when they see my build and discover my interests, assume I have a military background, but once they get to know me they realise nothing on this earth could be more anathema to my personality. Not just because the idea of killing people for a job is deeply unsettling to me, but more because I just couldn't deal with *anybody* telling me what to do. This is also the character trait that'll eventually drive me out of working in television, but that's another story.

For now, I was stuck in an office with a petty tyrant who held the future of my trip in his hands, so I just had to keep agreeing with him. Eventually, I figured, he would run out of patience.

'I am sure you are right, you know this area much better than I do. However, I must go to this area for my job, or my bosses back at the English tourist board will be angry with me. It is my job to go there.' A gamble – journalists are much mistrusted here – this was kill or cure.

'Just dirty *hutan*,' he snorted, 'you will see.' He stamped my *surat jalan* with a satisfying thump.

I threw off my sheet, and lay sweaty and sleepless. Despite so many years spent propelling my poor body from one side of the world to the other and back, I have never managed to conquer jet lag, and spend many long nights in the grip of its hateful insomnia; too tired to read, too excited to allow my brain to stop frantically playing and replaying nonsense scenarios like a bad inflight movie. None of this was helped by the fact that I'd just taken my antimalarial drug; a vicious pill called Lariam which has half a *Yellow Pages*-worth of potential side effects, including liver failure and psychosis. With me it just provided the joy of heartburn so evil it felt as if I'd swallowed a white-hot coal, plus vile twisted nightmares that made me wake up in a cold sweat believing I'd just eaten a maggot-ridden corpse, or murdered my own family with a chainsaw and buried them under a bush in the chicken run. I wake almost crying with the terrifying guilt that I'm a killer, and have to sit awake with the light on for several hours before the fear subsides.

This is so unutterably pathetic. I wonder if Alfred Russel Wallace took malaria pills. Actually, I know he didn't, as it was while he had malaria in Maluku that he came up with his theory that changed biology forever. And here am I, clasping my sheet to my chin because of a bunch of little tablets. Some great explorer *I* turned out to be! Finally I fell into a trippy, nightmare-ridden sleep in the early morning – just a few hours before the pre-dawn wails of the local mosque and competing cockerels made me wrap my pillow round my head and pray for unconsciousness.

My journeys have rarely if ever gone entirely to plan, right from my very first expedition. At perhaps six years old, my best friend Alistair Slater and I planned over many weeks to run away together on our bikes. I packed a bag and cycled off on my second-hand blue chopper bike, away on my big adventure, loudly singing the Brotherhood of Man's 'Running Away Together' as I pedalled into the sunrise. Just after dawn I turned up at Alistair's house, four miles away in the next town. I climbed up the drainpipe and knocked on his window as we'd arranged, but was met by an absolutely terrified face – it had never occurred to him I might actually follow through on our – to him – fantasy project. I remember the thrill of hiding under his bed while his mum grilled him about where I might be, and then my break for freedom riding off solo to who knows where.

The police picked me up a few miles down the road, but my mum and dad weren't cross. They'd been worried, but a little bit proud also. I think they knew even then there were certain parts of my character they'd never be able to quell – nor would they have wanted to. And I definitely blame my mother and father for my lifelong obsession with exploration, and with wildlife. We've lived our lives surrounded by animals, and it was inevitable that their passion would rub off on me.

They took my sister Jo and me on safari for the first time when I was about six and she four. Because we couldn't afford driving safaris we went on foot, camping out in tents and carrying big sticks in case the animals got too close. It sends a shiver down my spine even now when I remember the lions prowling round our tents and our black guide (unarmed) scrambling inside, eyes huge and terrified, to hide with us. Then later when I popped out to the loo, coming face to face with a grazing hippo right outside my tent, and the guide beating a charging hippo around the head with a paddle to stop it biting our boat in half.

After one memorable night spent sleeping on an open platform up a tree in the darkest bush, I lay awake on subsequent nights paralysed with terror, swearing I could feel the musty warmth of a leopard walking amongst us, feel it breathing and its great heart beating. The next morning at breakfast, my mother said, 'Oh God, I didn't sleep a wink last night – I was totally convinced there was a leopard up here with us', only for my sister and father to reply that they'd sensed exactly the same!

My mother and father are two of the most adventurous, smart and complete people I have ever met. If they had had the advantages in life that I've had, they would have achieved things far beyond anything I may have done. However, they were both born into working-class British families in which getting an education was not an option. They saved that privilege for my sister and me. Instead, both of them went to work for airlines at an early age, mostly because of the opportunities it offered them to travel. Dad worked for British Airways for over forty years, and for his efforts received several free flights for the family every year – an opportunity they made more of than you'd imagine possible. I have a vivid memory of our primary school geography teacher going round our class and pinning a world map with the places we'd been to. When the map was half-peppered with Backshall pins, she angrily told me to stop fibbing – the injustice still stings a quarter of a century later!

From the minute my sister Jo and I were born, the family travelled all over the world, but once we arrived in a place Mum and Dad were loath to spend any money, and actually seemed to take pride in doing things in as threadbare a manner as possible. So we lived like proper travellers. There is just no way I can ever thank my parents for the childhood they gave my sister and me. We had seen more of the world by the time we made our teens than most people will do in a lifetime, and they did it *properly*. No sitting about on a beach for us: instead, we went on massively over-ambitious walks, touring ruins or exploring woodlands or coastlines. We trawled souks and markets for bargains, hired rattling bikes or clapped-out boats, snorkelled the seas and yomped the hills of every continent bar Antarctica. Looking back, it seems that much of my early life was spent dragging a stuffed suitcase round Third World cities shrouded in the cloak of night-time, trying to find a room for under a pound a night, me with most of the bags because my sister was asleep over Dad's shoulder. I've been a professional traveller for fifteen years now and stayed in some utter hovels, but many of the roughest and scummiest places I've slept in were on family holidays as a kid. It certainly ensured that hard travel was second nature to me, and that nowhere was ever too much of a challenge.

Certainly not a valley in the centre of the world's largest jungle island.

Being as so much of Irian is just endless forest, I needed to concentrate my efforts in order to accomplish anything significant. The obvious place to start was the Baliem Valley, the only spot in Irian Jaya that really has any semblance of a tourist industry, and is home to some of the most celebrated indigenous cultures in the world. It looked as if it should be possible to head south out of the Valley, following the course of the Baliem River and wandering all the way into the Asmat region, an area of extraordinary remoteness and adventures the like of which I could only dream of.

With the city of Jayapura well and truly covered, I gratefully packed my rucksack and set off for the airstrip and a flight to the highlands. Passing through the lobby I tiptoed past the snoring hookers to avoid the inevitable fierce, lewd teasing they reserved for the clean young white boy. The airport was deserted, and after several hours wandering about looking for someone to tell me where my flight was, I had to hitch-hike back into town to talk to the airlines office. The air-conditioned office was staffed by a smart, portly Javanese man with a thin moustache and an unpleasantly false smile. It got even more unpleasant and false when I tried to book myself on a flight into the Baliem Valley.

'How are you going to Wamena, *sir*?' he asked with a sarcastic laugh. 'How are you going there when there is not plane? Will you walk to Wamena, *sir*?' He rolled his rrs in the characteristic Indonesian fashion, which I'd always considered rather charming, until now. There didn't seem to be an answer to this rhetorical question, so I maintained my tight-lipped politeness.

'Well, when do you think there might be a flight? I have to go there for my work.'

'Many people have their work, *sir*, but there are no flights. Did they not tell you about the burnings in England, *sir*?'

People in Indonesia are normally so deferentially polite, I was quite unaccustomed to this kind of treatment. I mean, you have to make some allowances for the peculiarities of translation, but this guy might as well have been thumbing his nose at me and flipping me the bird at the end of every sentence. I took a deep breath and forced myself to keep my cool.

'Of course I've heard about the burnings, but all the information back home said that it's mostly Sumatra, Borneo, Malaysia that's in trouble. People said Irian Jaya was fine – and the sky's clear here. Where does the problem start?'

My worst fears were being realised. The uncontrolled summer burnings that blighted an enormous swathe of Southeast Asia in 1997 were global news, with pictures of thousands in Kuala Lumpur and Singapore wandering the smog-ridden streets in face masks, choking their way through the thick, gloopy air. This had been one of my main concerns when coming out slap bang in the middle of the summer, but the truth is all my contacts had told me it wouldn't be a problem in Irian. Bad call.

'So what do I do? Do you have any idea when there might be a flight?'

The portly airline man gave me a haughty, harumphing laugh that simultaneously told me 'I don't care' and 'You're wasting my time and yours', before telling me to come back out to the airport the next day – never know, I might get a flight then.

Disconsolate, I took up my rucksack. I could already hear the cackles of the nightdress girls as I slunk back into their lobby. This was not going well.

2

In the seat next to me, a Papuan man wearing the distinctive cockerel-feather headdress of the Dani tribe, who barely reaches my chest when standing, pulls some tobacco from out of a pouch in the end of his penis gourd, which is poking out from under a grubby, moth-eaten Manchester United football shirt, and starts to roll up a cigarette. Trussed-up goats bleat from the seats behind me, and chickens dribble liquid shit into the aisles of the missionary plane. We are now hundreds of miles from the coast, right in the centre of Papua New Guinea's mountainous, forested spine, but as I look out of the window this seemingly eternal relief of dizzying remoteness and fecundity abruptly plunges away into an unexpected and remarkable landscape. Vertical limestone escarpments and spectacular cliff faces break from the green, and the tumbling scree slopes drop down to undulating, cultivated fields.

It had been another two days before I found a flight heading into the interior of Irian Jaya, and it wasn't on a scheduled plane, but a missionary airlines Twin Otter. Although all the travel writers I'd read who'd journeyed around New Guinea had used missionary airlines at some point, this was not something I'd seen myself doing; I just didn't think those airlines would be bothered with taking outsiders on their planes, and really didn't think it would be necessary. However, after two days journeying to and from the airport, the town airline office and my grotty hotel, I was willing to try anything. Not surprisingly, the MAF – Mission Aviation Fellowship – representative in Jayapura wasn't particularly wild about taking a travel writer into the Baliem Valley when they clearly had much more important things to drop off there, medicine and food taking up most of the flight's payload.

As I was kicking my heels in the MAF office waiting for information, a pilot came in, presumably to collect his flight plan. He wore classic aviator Ray-Bans, black tie and starched short-sleeved white shirt with gold-banded epaulettes – at a distance he looked more like an extra from *Top Gun* than a small-plane pilot in Papua. However, when I got close enough to introduce myself I saw that the jet-black hair scraped across his head with Brylcreem was inefficiently dyed and thinning, and probably the beginnings of a comb-over. He had a hefty paunch, was getting a bead of sweat on and looked totally bewildered – here was a man used to feeling the superiority of being an aviator, cast into a world that was just beyond his comprehension. His complexion was yellowy-grey and etched with liver spots and deep lines, and his dark chinos were polyester, shiny from being over-ironed down the years. His jowls wobbled in a slightly worried fashion as I shook his hand. He looked like a man who had clearly been meant to spend his days selling propane or running a hardware store back in Dakota or Minnesota, but had woken up in his late forties flying planes in some country he'd never heard of and was now doing his best impression of a master aviator, praying to God he'd not get exposed as a charlatan.

'Hi,' I said with a huge, warm 'please trust me and don't think I'm about to ask you for anything' kind of smile. 'My name's Steve, and I'm here writing a book on Irian.'

He blinked blankly beneath his Ray-Bans and looked for some means of escape.

'I'm trying to get into the Baliem Valley, and am not having any luck – you look like a man who might know what this weather's doing.'

He primped slightly at the mention of his credentials and the weather, a subject he as a pilot clearly fancied himself an expert on.

'It's this darn cloud that's messing with us, see, it's a real mess is what it is, I've not been able to put my plane down in Wamena for nigh on a week now.'

I made a sympathetic suffering face. 'Yeah, I know, it's a real pain. I've come right round the world to get into the Baliem. I was hoping to walk right down out of the Baliem Gorge.'

His eyebrows arched above his sunglasses in surprise. 'Oh see, that's gonna be real tough, don't ya know about the burnings?'

I sighed, deeply, my 'ah, the burnings, they've been my constant burden these past few weeks' sigh. 'Yes, of course. I kind of figured that a flight would get through eventually, though.'

'Well yeah, we're going to go right ahead and try to fly in today, but time's tough in the Baliem now. We're loaded up with drugs and rice and such. It's a regular famine in there, see?'

I didn't see. As far as I knew, the central highlands are known for being a place of plenty, with fertile soils where root crops flourish. People may not have material goods, but they have pigs and sweet potatoes in abundance, and never want for anything.

'A famine? What, from the burning?' I was struggling to see how the two things could be so potently correlated.

'No, it's your El Niño' he countered, surprised. 'They've had the worst drought in memory, the whole valley's as dry as a desert, see? Crops dying, pigs dying, no water or food. Reckon there's about five hundred people already dead from the starving.'

This had me stunned. Back then, the effects of El Niño and La Niña were known, but yet to be recognised as such powerful forces in global weather patterns. That year, however, was to be the moment El Niño made the world sit up and take notice, causing several thousand deaths and an estimated £20 billion worth of damage around the world.

'So what do I do?' I asked. 'I have to get into the Baliem for my job – is there any chance of getting on to one of your flights?'

'Well, like I say, we're going to go right ahead and try and fly in there today' – he was clearly not going to commit to anything – 'might even be more than one run if it all clears up top, but that's just up to the Big Guy, see?'

With this, he gestured up to the heavens, and the great Air Traffic Controller in the Sky.

It so happened that the Big Guy must have been looking down from the big control tower in the clouds, as mid-afternoon I was called up from my place sitting reading amongst the sacks of rice, and an hour later found myself next to the stocky Papuan in his penis gourd and football shirt, and dodging the green and brown chickens loosing shit in the aircraft aisles.

Travelling into the Baliem Valley from the north coast is truly one of the great plane journeys of the world. Around Jayapura the countryside is entirely deforested and the rolling hills look more like the Mendips than Melanesia, but just miles inland the landscape soars to peaks covered in rainforest. These are the Foja Mountains. With an estimated population of three hundred, they received no visitors at all till 1979, being too steep for conventional logging. This will change pretty soon with the advent of heli-logging. In one of the most dangerous jobs on the planet, heli-loggers are dangled into the forest, then huge choppers winch out the biggest and most valuable trees as they're cut from beneath. Given that a road is also being built between Jayapura and the Baliem Valley, these forests will soon be plundered – an absolute tragedy, as an international expedition there in 2005 discovered hundreds of new species of plants and animals. These included the wattled smoky honeyeater, a glorious bird with a red face and scarlet accoutrements dangling below its black bill. They also found twenty new species of frogs, and some of the largest new butterflies ever seen.

The tangled wonders of the Foja Mountains move beneath you for about an hour and a half, sublime forests over spiky mountains, as you head ever onwards into the dark heart of Irian. Just as you have set yourself to flying over jungles for a thousand miles till you meet the seas again, it's as if the mountains suddenly collapse in crumbling white limestone cliffs, dropping down to a plateau of rolling fields and farmland. This was very much the way that Richard Archbold came across the Baliem Valley back in 1938.

Archbold was the first Westerner to set eyes on the valley, when it miraculously appeared out of the rainforest as he flew overhead in his seaplane. All of a sudden, from flying over a vast wilderness of uncharted forests he found himself above a rural wonderland of ploughed fields, irrigated hillsides and settlements of beehive-shaped huts (known as *honai*), with their thick rush-mat roofs and chimneys sending occasional spires of smoke skywards. The sight was enough of a shock to *me*, even though I was expecting it, so it isn't hard to imagine Archbold's elation, even disbelief, at finding a forgotten civilisation on a high plateau, dead in the centre of the world's most remote jungle island. That would have been nothing, though, to the shock the fifty thousand occupants of the Valley must have felt. Having no idea that there was a world beyond the confines of their valley, they suddenly saw a huge metal bird splash-land on the lake in front of them, and a weird white man step out of its belly. It's easy to understand how such first contacts inspired 'cargo cults' to spring up throughout Melanesia – groups that believed all the white man's modern possessions and technology had been provided for them by supernatural powers, and who established religions to try and entice the 'cargo' to come to them.

The cultivated plain that spans the steep valley walls is, for the most part, a chess board of terraced fields, divided by rattan fences to keep pigs out and the crops segregated. All over the valley floor are jumbled assemblages of thatched *honai* in tiny settlements, like crops of mushrooms on a compost heap. A few crude, dusty roads and snaking streams carve up the plain, and the chocolate-coloured Baliem River slowly meanders across it before tumbling into fierce rapids in the southeastern Baliem Gorge. My plan involved making my way out of the Baliem Valley through the gorge, and following the river all the way down into the rainforest lowlands of the Asmat region to find the tribes of the Kombai and Korowai, some of the most extraordinary and fascinating tribal peoples in the world. It would involve a long and arduous trek southwards, perhaps taking over a month, and once I was committed there would be no hope of escape or retreat. It was a dive into very deep water, with little or no information to indicate whether my plan was realisable.

Our plane bounce-landed on the grass airstrip of Wamena, the sole real town in the highlands, which appears at first sight to be a blot on the wonderful landscape. Its rusty, brown tin-roofed buildings have deep, slurry-filled drainage trenches lining the streets, the only transport is clapped-out old minivan taxis, and heavy rains can quickly turn the lanes into rutted quagmires. But Wamena offers far more positive features than most Indonesian towns of its size. The streets are airy and spacious, as the valley people are used to living in small isolated communities and are loath to cram themselves together with hundreds of others. At a thousand metres above sea level, the town's climate is also excellent: cool enough to call for a blanket at night and rarely suffering the daytime swelter of New Guinea's sea-level towns. The missionaries have it easy here, their bright Wendy houses gleaming behind white picket fences that enclose springy, manicured lawns, and immaculate blond-haired kids to be glimpsed at the windows. These Little Tennessee havens seem peculiarly incongruous among the naked natives and their ever-present pigs.

Wamena is a place where the sheer weirdness of Irian Jaya is at its most stark. The Indonesian occupiers have built a mosque here, with a shiny minaret. Women can be seen walking the streets wearing burkhas with only a narrow slit for the eyes, and are affronted by any immodesty when it comes to dress. Additionally, as is well known, Muslims abhor pork and anything to do with it. Then you've got the native Dani, wandering around stark-naked except for their whopping penis gourds, whose entire monetary and cultural system is based on the pig. Why the two don't live in perfect harmony is just beyond me ...

The indigenous Dani people have been the subject of fascination for global anthropologists ever since Archbold first encountered them and brought back tales of a people still living Stone Age lives, unaware of the outside world. Until very recently, the Dani chopped the Valley's splendid casuarina trees – named after the cassowary, as the trees' foliage resembles the bird's plumage – with axes made from chiselled malachite, and even now the tribesmen still squat on their haunches smoking, utterly naked but for a penis gourd even during the Valley's freezing-cold evenings.

The Dani male's groinal apparel is still omnipresent: it's called a *horim* or *koteka*, and consists solely of a gourd which encloses the penis shaft and points upwards in a permanent erection. Like the Elizabethan codpiece, its function is to represent and accentuate the engorged male genitals, and assert masculine power and fertility. The different tribes of the Valley differ as to whether they use a long thin gourd or a short fat one – sometimes it pokes out from beneath a hooped wooden skirt. It's tied at the bottom around a gathered flap of loose scrotum skin or around one testicle, and at the top to a thread that encircles the waist. With some *horim*, the man needs to actually invert his penis into his body in order to properly wear it (which makes my eyes water just thinking about it). The *horim* is most definitely the piece of clothing you would least like to get caught in an elevator door. Thankfully, the day of the electric lift has not yet dawned in Wamena.

Unfortunately, Irian Jaya is catching up with the rest of Southeast Asia on at least one front, and that's the burning of its forests to make way for arable land and, increasingly, oil palm plantations. This tragic phenomenon accelerated rapidly in the latter half of the 1990s, and the scale of the now annual disaster is just impossible to comprehend. The summer of 1997 was the first time Asian countries actually started calling states of emergency to deal with the debilitating carcinogenic smog. And it wasn't just the burnings that were blighting lives. That same summer saw the beginning of one of the harshest periods in the memory of the peoples of Irian's central highlands. The El Niño weather system had terrible effects all over New Guinea. The eastern half had to receive supply drops from Australia, and much of Irian Jaya's normally drenched forests dried up. Fires started by slash-and-burn farmers raged out of control, and locals started more fires because they thought the smoke in the sky would become clouds and cause rain. Missionary mercy flights were also grounded, and by the end of 1997, after four rainless months in the usually lush Valley, over five hundred people had starved to death in the immediate area of Wamena.

If I had had any inkling of how bad the problem was, I would have stayed well away, but you can't get news between neighbouring villages in Irian, let alone to the outside world. Having waited so long in sweaty Jayapura for a flight, I wasn't about to get out of the plane at Wamena airstrip, discover the country was all a bit parched, then fly straight back out again. Besides, the scale of the problem was not really evident there in Wamena; that was yet to be revealed as my journey proper began.

The idea of following the course of the Baliem River certainly seemed to have merit, pursuing it out through the tumultuous gorge that the watercourse has carved as it forces its way out of the highlands. Beyond there, the river drops several thousand metres, cascading down to the jungles of the Asmat. This remains one of the least known but most exciting places left on the planet, with uncontacted peoples living in totally unexplored forests, packed with birds of paradise and other natural wonders.

I had my heart set on the Asmat, and more specifically the Kombai and Korowai regions, as I had a tatty old *National Geographic* article on the people that lived there, inhabiting rickety tree-houses twenty or even more metres up in the forest canopy. Their lives were to anthropologists the Holy Grail: a living culture that has remained oblivious to the rampaging bulldozer of world history for several thousand years, and exists as a living connection to our own hunter-gatherer origins. The *National Geographic* journalist had found villages that had never before seen outsiders, and been greeted by drawn bows and brandished daggers. It looked to be a journey of several hundred miles, much of which I would have to do on foot. Information was impossible to come by outside of Irian Jaya, so my first week in the Baliem Valley was spent on research. I spoke with all the people in Wamena who might know anything about the area, before wandering out to the smaller villages that were easily accessible from there, in search of information. I ended up sitting in on tribal ceremonies that were obviously arranged solely for the tourists – that is, me – as well as inspecting one of the only mummified corpses left in the Baliem.

Not so long ago, all tribal Big Men would be mummified in order to preserve their mighty spirits, by being bound in the sitting position, knees clasped to chest. The corpse would be drained of fluids and left to hang in a tree. But outraged Protestant missionaries torched several thousand of these 'fetishes' in the 1960s, and outlawed the practice. It's hard to imagine how the locals must have felt, seeing their sacred ancestors thrown on a bonfire. The mummy I saw was one of the very few remaining. With its permanently open yelling mouth, the black shiny leather of the flesh drawn tight over its bones, it looked like Edvard Munch's painting *The Scream*.

On a street corner I saw vegetables being sold by an old woman caked from head to toe in dry, cracked grey mud – the symbol that she was mourning a lost husband. Many of the women were missing digits, as custom dictates that grief should be expressed by cutting off one's fingers with a bamboo knife. Apparently the only anaesthetic offered before the fingers are sliced off is a dead arm, cheerfully administered by a punch from a willing elder! The women here all wear woven bags called *bilums*, with the straps over their heads and the bag part dragging down their backs. They keep everything in them, from vegetables and packages of fish and prawns to their dogs, piglets and babies. The *bilum* is of immense cultural significance, evident in the name: *bilum* means 'womb'.

Also on display in the market was the sad evidence of Wamena's bushmeat trade; one fellow tried to sell me a long-beaked echidna, trussed up like a baby in a woven palm blanket, but dead and bound for the cooking-pot. These incredibly rare creatures have to be one of the most bizarre animals on earth. Along with the platypus, the two species of echidna are monotremes, egg-laying mammals that suckle their young after they hatch with milk sweated on to the skin rather than through nipples. As if this isn't weird enough, the male echidna owns one of the weirdest sexual organs found amongst mammals. Four-pronged and about ten centimetres long when engorged, it looks something like a drill bit!

Although I've seen the short-beaked echidna many times in the wild in Australia, I am still to see a long-beaked alive, and it seems unlikely now that I ever will; they are locally extinct throughout much of New Guinea. Another stall-keeper proudly displayed a brace of dark cuscus, soon to be forming the basis of a highland stew. These possums are also marsupials, reflecting the distinctly Australasian fauna found throughout New Guinea.

During the research period, I streamlined my rucksack and bought gifts for the tribes that would have to be charmed en route. Though I was getting quite a flavour of the Baliem Valley itself, I was pretty much drawing a blank on what might lie ahead once I made it out of the gorge. In Wamena I stayed in a little homestead across the grass airstrip from the town. Every evening they'd cook me fire-blackened sweet potatoes and swollen prawns from the Baliem River. Though they tasted buttery and sweet, they looked for all the world like great brown cockroaches, and it was quite an effort forcing the meat down. But ahead of me lay uncertainty, and meals such as this would soon be the stuff of fantasy, so I scoffed the bugs down, trying not to think of their crunchy insect lookalikes.

I spent a week in Wamena, gathering as much information as existed from some very suspicious characters, all offering to guide me to the Asmat region for about the same price as a small house, only to tell me it couldn't be done when I said I'd be trekking alone. The night before departure, I made my way to the only telephone exchange in the town and called my parents. They sounded a million miles away down the crackly line, and I tried not to sound worried or to give them any indication of what I might be getting myself into.

'*Stephen?*' – my mum is the only person who has ever called me this, and even she saves it for when she's cross, or excitedly answering the phone to me – 'Stephen, yes, we can hear you! Where are you, what are you doing?'

Over the years, I've learnt that honesty is not always the best policy when telling my folks what I'm up to. There is nothing they can do, and they'll only lose sleep and worry over my safety. All the same, that night I had a fierce urge to tell them everything – to hear how proud they were of their independent, adventurous son, to calm my mum as she cried down the phone. Instead, I just told them there'd be no phones for a while so they shouldn't worry, and that I was just going for a bit of a wander in the gorge. Hanging up, though, I felt thoroughly alone and a bit miserable. I really needed a hug.

My trek into the Stone Age was to begin in the southeast of the Valley, where the fields and furrows end and the magnificent untamed Baliem Gorge begins. Here, the river speeds up to become rapids, tearing violently between the steep canyon walls. Looking down at the tumult of the river here, I vowed one day to come back with my kayak and try and run its whole length from source to sea. Though I'd managed to get data and a small sketch-map of the villages that peppered the banks of the river as it dropped out of the highlands, I was still drawing a blank on what might lie beyond. It seemed to me that the best thing to do was just to get going and hope that, as I pushed further on, people would have better information. As soon as I was beyond the reaches of the Valley, walking would be the only way for everyone to move themselves and all their goods around. They surely would know ages-old trade paths that I'd be able to follow.

The first steps of a long journey off into the unknown, carrying your world on your back, are always indescribably exhilarating, ecstatically empowering. You feel as if you could sprint for a thousand miles, whooping and yelling: 'I am slave to no man, the path is as I make it, I move as I will and sleep where I fall!' It is the ultimate existential joy. Curiously, though, this is probably a cultural rather than a universal sensation. In Indonesia, whenever I encounter local people and they find I am travelling on my own, they are first mystified, then sympathetic. Their culture is based around collusion and community, people are only ever alone if they are outcasts. This is obviously a generalisation, but whereas two Westerners will come to a deserted beach and take opposite sides of it, keen to avoid each other, not to intrude, when many Asian peoples find themselves on the same beach they will instinctively come together and share the same spot, crowded sociably together while the rest of the beach stays empty.

When asked by locals if I am travelling with friends or family, I proudly respond: '*Tidak, saya sendiri*' – 'No, I travel solo.' What I cannot communicate is that for me this is a good thing. In the local tongue, *sendiri* means not only solo and solitude, but single, solitary, lonesome, alone. I mean to boast that they are witnessing a free spirit in control of his own destiny, but what the locals hear is: 'I travel lonely.'

I have always valued my own space, and spent a lot of time on my own. As a child that meant long exploratory runs with our tireless black mongrel Buster at my heels, building camps out of bracken in the woods, or rope swings up by the foxes' earths, trying to make fire by rubbing sticks together, or setting snares to catch rabbits. On family holidays I would quite happily wander off all day, nosing in rockpools, exploring coastlines, catching lizards, and I'd infinitely prefer my own company to trying to make friends with other kids. After all, I might not have succeeded and that would have been shameful. Amongst my heroes are people like Alfred Russel Wallace, whose ability to just dump himself in nineteenth-century Borneo and be totally self-sufficient – one minute declaring a new species of primate, the next bargaining with cannibals for his life – I admire tremendously. There's also a liberating purity about adventures in which someone just sets out on their own, wandering off into the wilderness, like Laurie Lee in his wondrously romantic book *As I Walked Out One Midsummer Morning*, or Keroac in *On the Road*. Or the Littlest Hobo, the dog from the eighties TV series, for whom 'there's a voice, keeps on calling me, down the road that's where I'll always be. Every step I take, I make a new friend, can't stay for long, just turn around and I'm gone again. Maybe tomorrow, I'll wanna settle down, until tomorrow, I'll just keep moving on, until tomorrow, the whole world is my home!'

There's so much romance to that idea of just going, with no great plan and no real fixed idea of where you'll end up. Under your own steam is important too: I've never had any time for people who drive across the Sahara in a four-by-four or take a skidoo to the North Pole and call it adventure. That's not adventure, that's not testing yourself to the physical limit – that's going for a drive!

I can remember as a kid being glued to a pulp Australian minidrama called *The Shiralee*, starring Bryan Brown as a gnarled outback stockman who just wandered Australia between jobs. He had no fixed abode, merely his swag (sleeping mat) slung over his shoulder, and slept where he felt like it, ate when he had money or could catch something. I totally fell in love with the idea, and still, even now, see that as the ideal life for me. Mortgages, jobs, commitments, money and possessions have got in the way, and perhaps will always tempt me back, but really I know I'm happiest, most fulfilled, when I wander lonely.

3

Early morning, and a boiling mist seethes through the Baliem Gorge, framed by craggy mountains steadily more hot-wash-faded with distance. The steep sides of the gorge, planted thick with sweet potatoes, cascade down to the rampaging Baliem River, its constant rumble like a distant jumbo jet warming up its engines. The sun breaks above the peaks, the mist burns off like a startled spirit and the whole character of the gorge changes: a yellow light now warms the bedewed hillsides, my sweat-soaked T-shirt starts to steam, and birdsong begins to reverberate round the craggy arena, drowning out the thump of my heart in my ears.

My initial excitement to finally be under way, that my grand adventure was becoming a reality, was tempered by the certainty that to get to the trail itself would *not* be one of the world's great journeys. I negotiated the road south from Wamena in the dying shell of a rattle-box minibus, over roads as smooth as the average open-shaft mine. All of the minibuses that ply their trade across the valley have come here after reaching the end of their working life back in central Indonesia. They look as if they've been dropped out of a passing jumbo jet, battered and rusting, held together with chewing gum and prayers.

The scenery outside the windows, however, could have been from *Little House on the Prairie*, with its neatly divided fields and children waving and screaming in unison as we passed. Inside the minibus, though, was a sardine sauna without the prospect of a bracing bath to follow. Most Indonesians believe that open windows in a moving vehicle will give you a head cold, so keep them firmly shut, even when it's thirty-five degrees outside. The minuscule gap between my feet was occupied by another brace of traumatised chickens, obviously fully aware that they were only still alive because of a shortage of Tupperware and that upon arrival they'd be packed lunch. I couldn't therefore blame them for the green poo – yet more of the stuff – trickling all over my boots.

The thirty or so passengers inside the minibus (known here as a *bemo*) were all seated upon two giant industrial speakers that ran the length of the vehicle and could comfortably have provided the sound for a rave in an aircraft hangar. The music of choice was Schlock Asian pops, and the bass was turned up so loud that the bemo bucked up and down even when at a standstill. I had spent so many hours on similar buses listening to the same ubiquitous tape, and my patience was close to expiring. Crammed right at the back of the bus, I spied a wire coming from beneath the torn lino floor and running under the speaker. After half an hour of fantasising – should I or shouldn't I? – then cajoling myself, I surreptitiously took out my trusty penknife and opened the scissors. Bending down as if to tie up my shoe-laces, and what I was up to hidden by the shivering chickens between my feet, I found the wire and snipped it.

Nothing happened. If anything, the noise appeared to get challengingly louder. After a suitable guilty wait, I bent down amongst the rooster crap and the clove cigarette butts and found the wire. I pulled on it gently, and from under the bit of torn lino it unravelled, and led straight to the rear brake light.

Several hours later, the bemo sputtered to the end of what passes for a road at the administrative centre of Kurima, a small, ugly village on the brink of the gorge. Local people shouted myriad instructions at me, and tried to separate me from both my luggage and my sanity, to drag me off to see the corpse of their grandfather or buy some tatty carving from their aunt. But I ignored them, plucked my pack from the roof, and staggered off up the well worn and extremely steep path that led straight up the mountainside, looking and feeling under foot much like the bed of a dried-up waterfall. There's an airstrip and a mission at Kurima, both cut precariously into a precipitous rockface, three-hundred-odd metres above the valley floor.

As a spotty eighteen-year-old on my very first solo travelling jaunt, I'd waved goodbye to my sobbing parents from underneath a backpack the size of a Volkswagen, containing all the useless travel tat well-meaning relatives had given me as birthday presents – including a baseball cap with a fan in it, for God's sake! But over the years travelling in this part of the world I'd managed to pare the pack down to the true essentials. I wanted to do this all for real, and solo.

My pack contained tobacco and betel (a blend of crushed lime, betel leaves and the gnarled nut of the areca palm, which is chewed for its mild narcotic effects and leads to red staining of the lips, teeth and gums and requires the chewer to spit bloody gobbits into the dust) for gifts, sleeping bag and mat, a warm change of clothes for the night-times, water and med kit. I'd not brought any food or shelter, as I'd learnt that even in the furthest-flung corners of Asia people welcomed you into their houses with open arms. In most small villages you first turn up and meet with the village head, share some tobacco and betel, and they will provide you with a place to stay and meals with the family. It's always simple fare – maybe just some rice and rock salt – but it enables travellers to spread their money amongst those who need it and to meet and learn about the real local people, not to mention saving the effort of carrying huge amounts of food on your back.

I was determined not to do the tourist thing and hire a guide to lead me or a porter to carry my backpack, so I plodded up the mountainside in the raging heat, followed by a huge grunting pig who seemed to be enticed by my discarded banana skins. My experience had also shown me that you don't actually need a guide, as with a little of the local lingo you can just keep asking for directions. However, as I clambered up, every twenty or thirty metres the main path would branch off into several promising-looking mini-tracks leading to God alone knows where. On the fourth or fifth occasion of finding that the trail I'd chosen led to the front door of someone's little beehive house, and having to excuse myself to the bewildered-looking occupants, it became clear my confidence had been misplaced.

In his book *Into the Crocodile's Nest*, written a decade or so earlier, explorer Benedict Allen reported having flown past the stretch of gorge I was heading into because unfriendly, armed natives rendered it too dangerous. Things had changed since he was here; certainly no one had warned me against travelling through Tangma, and it didn't feel at all threatening. I was being a little naughty, though, not having declared this part of my journey when getting my *surat jalan* stamped in Wamena's police station. The way I saw it, they'd probably just have lumbered me with a police guard, and I could always just plead foreign ignorance if I encountered any military on my way. Certainly there would be no police stations once outside of the gorge.

As Kurima disappeared behind me and I reached the upper slopes of the gorge, all of a sudden the scenery deteriorated unpleasantly. It was like walking through the warpath of a recent lava flow, all the vegetation around me smouldering, and the back of my throat stung with the acid rasp of smoke. Even the earth was black and singed, and the surrounding shrubs charred into crusty black skeletons. The air was filled with the constant hollow thock of axes on the few remaining trees. The pig still following tight at my heels, every so often I would meet the perpetrators of the destruction, as groups of men in extravagant headdresses and penis gourds would loom out of the gloom, straddling the tree-tops hacking down branches, or putting flaming torches to the thirsty undergrowth. Every time a tree fell, the men would set up a whooping chorus of delight.

This carnage soon evaporated once I stepped over the 2,000-plus-metre col between the Baliem Valley walls and looked down into the neighbouring valley. The mission town of Tangma snuggled below me in a steep gully, shaped dramatically by the shallow scar of a river. The adjacent slopes were cut into neat steps for the cultivation of corn and the staple sweet potato, and I noted the native huts by the pathside. As I descended, their inhabitants came out to greet me, and to laugh and point at the pig still following at my ankles as if it had been instructed to 'Heel!' After having trekked for several hours up forbidding cliffs, with no views to recompense me, this sudden glimpse of the greener grass on the other side of the valley slopes revitalised and sharpened my senses. Just as well, really, as the scramble down was strictly for mountain goats only. My breathless misery was compounded as I was regularly overtaken by normally breathing locals, many of whom looked to be in their eighties – and every one with calf muscles like rugby balls and carrying huge bundles of vegetables, children or massive grunting pigs tossed nonchalantly over their shoulders. Somewhat typically, though, it was the women who bore the heaviest loads, with the men obviously far too busy smoking tobacco and spitting.

At one of my interminable enforced detours, I was beckoned inside a hut by the men who resided there, the women being banished to their own separate quarters. The doorway was no more than waist height, so required me to go down on all fours in order to get inside, trying not to cough from the acrid smoke as my eyes got used to the darkness. I waited patiently for a cup of tea or some other offering, which never came, and they seemed to be waiting patiently for me to perform some kind of magic trick, staring at me with bright, expectant eyes. Inside it was as black as pitch, the wooden stakes that formed the walls coated with a thick tar from the constantly smoking fires within. The smoke leaking up through the thatched roof made it seem as if the whole hut was on fire. The walls were lined with benches that for now made seats, but would double up as beds for the men at night. I asked how many people slept in the hut, and a hurried calculation came up with a figure of fourteen. It was no more than five strides long by three across.

Not so long ago, the town that lay below me was out of bounds to anyone without a large gun. I had, however, been assured that Tangma was now as safe as it looked. The little town is arranged around a rarely used grass airstrip, lined with wooden mission chalets. The Irianese seemed to have moved out, and had built new *honai* on the outskirts, well away from the twee new houses. They wore scrappy T-shirts and shorts, and there was little evidence of the cultural idyll I'd come to see, so I decided to push on through to the areas where the scenery really starts to get special. As the afternoon wore on, I passed waterfalls tumbling down the steep gorge and the shallower slopes full of bare-chested women, harvesting enormous vegetables that would cause a furore at any village fête back home in England. Higher up, the pandanus trees stood like something out of Dr Seuss, with their expansive palm crowns and huge pineapple-like red fruits. The Dani value these oily nuts only slightly less than their pigs, and when they're in season they organise festivals around their consumption.

I sat for a rest on a curve in the path that offered a staggering view of the Valley. Several of its bustling inhabitants were at work on the slopes. I had a couple of (green) oranges in my bag that I'd picked up at Kurima, and started to peel one. My pig came over to join me as bold as brass, and waddled right up between my feet like a great big black dog – albeit one with a serious outbreak of dermatitis, his coarse skin and thick bristles a wire brush scouring my bare legs. He snuffled at the proffered orange, before taking it gently and scoffing it down.

'I shall call you Shep,' I told him solemnly.

Pigs in the Baliem Valley are not a customary foodstuff, being far too valuable, and are only really eaten at big feasts. The highlanders have a very distinctive diet, dominated by sweet potato but supplemented by fruits such as pandanus and *buah merah* ('red fruit'). The latter is a large spiky pod, like a sadomasochist's loofah. The insides are crushed and then boiled, and taste a bit like dark chocolate.

Around me on the cliffs the Dani women were bent double grubbing up sweet potatoes, their knee-length grass skirts looking remarkably uncomfortable sitting just below their hips. All women are considered to be witches in the highlands, possessing a powerful magic that increases with age. Old women can put curses on men, causing them to become infertile and sometimes die horrible deaths. This doesn't actually seem to bother the Dani men that much, who spend most of their time engaged in the far more noble art of smoking, some wearing headdresses made of cockerel feathers arranged in a fetching and photogenic crown, often with longer, more elaborate feathers falling down to frame the face. The headpieces always used to be made from bird-of-paradise feathers and cuscus fur, but now these vibrantly coloured creatures are all but extinct. You still see their feathers and fur in antique headdresses, though dusty and sad, but the feathers still retain something of the insane colours of the birds beyond beauty that they once adorned.

As I continued south, I had to cross the raging Baliem River on heart-stopping suspension bridges, their flimsy-looking planks sometimes bound with steel wires and sometimes with meagre palm ropes. During the height of the rains, when the river is in full force, the waters actually lap about the boards because each step makes them lunge downwards. In one instance, next to the bridge I was about to face were the sorry remnants of the old one that had collapsed a few years back, its skeleton hanging there to inject a touch of apprehension into paranoid hikers.

This wasn't my first such bridge, and I strode on to the shaky slats, but Shep wasn't as confident. He tried out the first one a couple of times, sniffed at the wood, then backed off, obviously unsure.

'Come on Shep,' I shouted, 'it's perfectly safe!' He looked up at me, had another sniff, and then turned around and started slowly back up the path.

'Shep!' I called desperately. 'Trust me, it's fine!', and to demonstrate its solidity I jumped up and down on the boards.

It was then that I had a moment of piercing clarity. Jumping up and down on a Papuan rope bridge to prove it was safe – to a pig. The bridge was too much for my travelling companion, who turned and trotted back in the direction we'd come from. I never saw him again.

The walking could have been an absolute joy, the altitude taking the edge off the heat, the views divine enough to still the grumbling in my stomach. It could have been paradise, had it not been for the attention of a few – impossibly hardy – flies. These were tabanid flies, also known as gad-, deer-, horse- or cleg-flies. Their sharp saw-shaped mouthparts slice open their prey so they can lap up the blood. However, while your average horsefly will split open like a tiny melon if you're quick enough to swat it in the act, these evil little blighters were totally indestructible. I nicknamed them 'robobugs', as they seemed to actively enjoy insect repellant, barely bothered flying away when you slapped at them, and would respond to a hearty open-hand whack landing on target by shaking themselves down, taking a deep breath, then nailing you again. Anyone who's been bitten by a horsefly can attest that they really, *really* hurt. After I'd been bitten three or four times I found myself sprinting away, arms windmilling blindly like a ten-year-old in a playground brawl. Eventually I had to forgo the pleasure of walking in shorts and T-shirt, and put on my trousers and shirt, which turned me into a sweaty mess in minutes.

As the dark closed in, I came upon a cluster of *honai* without the – by now customary – missionary station nearby, and decided to see if they'd help me out with somewhere to sleep. My reward was a display of bright, white grins and a stool outside one of the huts, while a youngster headed off to get me some tea. Dinner, though, was another proposition. In my years trekking in Asia I've very rarely had any difficulty in this department: with a little of the local lingo, some cash and tobacco, you can usually get yourself a meal and a bed just about anywhere. However, I'd reckoned without the shortage of food here. Despite the vegetables I'd seen people collecting on the trip in, it seemed food was in very short supply, and the folk had little or nothing to give. This left me in a terrible position: take food from their children's mouths and pay them money they couldn't use, or continue trekking on my already rumbling and hunger-weak stomach. I decided on the latter.

I spent that night in a thatched Dani hut with soot-blackened, greasy walls, wrapped tight in my sleeping bag on a bench bed shared with about ten other men. Pigs sleep in a sectioned-off part of the *honai* with their owners, and are such valued members of the family that if a sow dies it's customary for the piglets to be suckled by a woman of the household. I kind of figured this must be some urban myth, until the first time I saw it happening – a woman sat outside her hut with a hungry piglet clutched to her breast, pugnaciously nudging away with his nose. It looked not only incongruous, but downright painful.

My companions were all male, and to begin with I just sat there, obviously a guest-of-honour they had no idea what to do with. After a little while, I felt a tell-tale rumble in my stomach and an uncomfortable build-up of gas. I remember as a child my parents telling me that there were certain cultures around the world where a resonant belch was considered the best of manners, but was pretty sure that letting rip was not a Dani gesture of goodwill. But I was bursting, so I made a snap judgement as to my company and the type of fart that was likely to come out, and decided it wasn't going to be too noisy and I'd probably get away with it. So I quietly squeezed it out.

Unfortunately the early mornings, sweaty hours of walking and meagre diet had obviously taken their toll on my digestive system. The resultant expulsion was indeed silent, but spread in an acrid miasma of foul gas you could practically see – even taste. As it drifted through the hut, I sat there squirming with embarrassment. The guy next to me had small smiley eyes, a thin curly pubic bush of a beard and a couple of small blossoms in his hair – he'd pushed a biro through the hole in his nose (where you'd expect a bone to go!) earlier that afternoon, and had a mischievous air about him. He was idly rolling tobacco into a dry leaf when the smell hit him. His face wrinkled in utter disgust, almost like a bad actor doing his 'just smelt something rancid' face. Then came the wave reaction from the others around the room. After a couple of seconds of silence, I held my hands up, open-palmed, and said: '*Tidak saya*' ('It wasn't me'). The guys folded up. They slapped my legs and gripped my arms, breaking down all our awkwardness. Their hysterics lasted for a good ten minutes. It was the first time I'd made friends by farting in someone's living room.

We sat snuggled around a hypnotising fire, chatting idly and laughing about the same subjects that any group of men around the world get round to when they find themselves together: sex and – er – pigs.

Well, I didn't have much to offer on the latter subject, but I had plenty of opinions on the former. The men ranged from two youngsters in their mid-teens through to several wizened grandfathers whose mischievous faces broke in seconds from sullen concentration on rolling a cigarette into brilliant toothy grins.

'So, Mister Istiv' – nobody out here could quite master my name – 'why you come to Baliem? Are you from the missionaries?' The group leaned forward to hear my response.

'No no! I'm a writer.' This simple answer met with a lot of discussion. It's quite a difficult one – the Indonesian word for 'author' often draws very blank stares. 'Journalist', I try, with the accompanying scribbling gesture that is ubiquitous in restaurants when you want the bill.

'Oh! Journalist!' they nod, as if in total understanding, but the eyes are still not comprehending, and they toss the idea and the word around between them as if trying out its efficacy.

'You come here to buy something?' asks one man.

'No no, I just come here to see, to see the mountains and the forests and the river.' I stop short of saying anything about seeing the people and how they live. Somehow it would seem weird, as if this were a zoo and they were the animals to be gawped at.

'It's very beautiful here, I see places, write about them, and then my friends in my country will want to come here too and see how beautiful it is.'

'So your country, how long does it take to walk from there?' an older man with rheumy eyes, a greying beard and a ring of cowrie shells (how did they get here?) around his neck asks. This throws me rather. It strikes me as a question I probably should know the answer to, but don't actually have a clue.

'Well, I didn't walk, I flew here in a big plane.' I make a child's aeroplane with my arms and mimic the noise. They nod – I'm sure they've seen planes coming in to mission airstrips. 'But I guess if you were to try and get here walking, it would take a year – no, maybe even two years.'

This gets a big reaction, some tut-tut-tuts and a shaking of heads. Actually, maybe it would take three years. Maybe I should find out!

One serious-looking young man who is clearly of quite a proud nature thumps his chest. 'I will walk to your country. This is not far for me.'

'Of course,' I venture, on dangerous territory here. 'And my house will be your house!'

What a marvellous image he conjures up! What would modern England make of this young man, naked even in the cold highland night except for his penis gourd adorned with cockerel feathers and bits of moss, yellow and white orchids in his hair and a dagger tied to his calf made from a cassowary shin bone. Actually, thinking of how lost this strong, noble young man would be in our alien world gives me a momentary pang of sadness. How could I explain to him that in my world the most important thing is sport? That the only thing our country will mobilise into frenzied fervour over is football. That people care more about the activities of 'celebrities' than they do about their own family and neighbours.

In the pub back in my own village one day last summer I met a young girl. It was hot and she was wearing a woolly hat pulled down low. What was all that about?

'Oh, I've just had my botox injections, and some filler put in my lips. My dad got them for my birthday,' she replied.

'Really? How old are you?'

'Twenty-one.'

How would I begin to explain my bulimic, brand-named, self-obsessed world to my Dani friend with his grand continent-striding intentions? How ludicrous the things Westerners take seriously are! How wasteful we are – my weekly rubbish bins would probably contain more of value than he would ever own.

Luckily he's now on to more important topics: 'Mr Istiv, in your country what are the women like?'

Hmmm, now there's a question I could talk about at length.

My new friend continues: 'I saw a white woman in Wamena once, and she was like a boy!' He wrinkles his face in an expression of contemptuous distaste, while a man slightly older mimes a body shape for him – he shows me a girl with the figure of Christopher Robin. Ah, so that's what he means, the all-important subject of breast size. I muse on how best to address the topic, knowing that out here the pendulous swinging octopus head is very much the bosom of choice.

'Well, women from my country have many different sizes of *susu*,' I say. *Susu* means 'milk', and seems to be the best word on offer, though it does make it sound as if I'm talking about udders. 'Some girls are thin and like you say, but others are – well—'

The first props that come to hand are my rolled-up sleeping bag and my cooking pot. Stuffed up my jumper, they create a huge lopsided rack that causes unbridled mirth lasting for what seems like hours. I wish my mates back home thought I was this funny. They slap my legs in merriment, as if I've been a very naughty boy.

'Mr Istiv,' asks the rheumy-eyed one pointing up to the sky, the group hanging on for my answer with drawn breath, 'have you been to the moon?'

I set off the next morning, waving goodbye to my friendly penis-gourd-clad hosts. Requisitioning one of their children to lead me on to the next village, Werima, I headed south along the banks of the Baliem.

Early in the mornings, a tissue of mist lingers over the river before the sun scorches it away and makes a hiker long for the bone-numbing cold of the highland evenings. The scenery along this route is dramatic: great ravaged cliff faces torn by landslides and earthquakes, and delicate waterfalls. Scree-covered slopes tumble down to the Baliem and its tributaries, dark-green vegetation clinging desperately to any plausible holds. Breakfast had consisted of a tart orange and some dry crackers, and did nothing to ease my ever-growing hunger or my concern about where the next meal might come from. As I tramped on through the morning, which was getting warmer all the time, it felt like I was taking myself, step by step, further away from food. I'm quite a hefty fellow, with a fierce metabolism. At university my day would start with an entire tube of Weetabix and a protein shake, and I'd be hungry again by about eleven. While I was no longer quite such a glutton, my calorie intake since leaving Wamena was less than I would usually consume in a single midnight snack, and the walking was quite hard. My stomach growled at me grumpily, cursing me for not packing at least a big bag of rice.

The footpath heading towards the jungles of the south passes through the airstrips and missions of Passema and Soba. They were set up in the sixties and both towns are centred on the houses around the airstrips, built more like Swiss log cottages than Baliem huts. The dark-planked walls are surrounded by pristine flowerbeds, their windows and gables edged in gay purple and blue paint. It must be quite an experience to land here, these 'airstrips' at first sight little more than small, grassy village football pitches at impossible gradients and about as flat as a mogul ski slope.

The people here are far from used to Western faces, and I was led hand in hand with delighted local schoolchildren to the house of the mission schoolmaster, who keeps a room set aside for unexpected tourist guests. A young girl, perhaps twelve years old, with her head shaved, and naked except for a pair of tatty knickers, was rolling an old bicycle wheel without the tyre and most of its spokes over the dry ground, using a stick to keep it moving. Her younger brother, with gallons of snot pouring down his face, and flies in his eyes, sat with his huge distended stomach, crying in the dust. A fire was burning in an old upturned oil drum, the kind of device you expect to see Pittsburgh tramps warming their hands around.

Otherwise, the village was deserted. Perhaps the women were out in the fields, and the men busy chopping trees and starting fires. With the smog blocking out the sun and casting everything in its sallow grey light, it felt like I'd landed in some post-apocalyptic wasteland. Sitting down on a log, I put my head in my hands and pondered my predicament. This was not the Shangri La of the highlands I'd been expecting. I had blundered in on their hard times on a pointless, selfish mission, and felt ashamed of myself and my intentions. In the schoolmaster's room, a set of boards covered with a few rough blankets proved infinitely preferable to the attentions of the previous night's bloodsucking bedfellows, but still offered precious little sleep;

particularly as the noisy scrabblings of rats in the rafters got progressively closer and closer, until at one point I opened my eyes to see a rat eating my foam pillow, just inches from my face.

The missionaries are pretty much omnipresent in the central highlands. There is actually a map in the missionary airlines office that has the whole area painted like a multicoloured patchwork quilt, even the most wildly remote places, with each colour denoting not native boundaries or the people who live there, but which Christian sect has claimed and converted the locals – the most obscure American denominations seeking to stamp their own brand of cultural imperialism on the world's last 'savages'. Many agnostic and atheist travellers find it truly objectionable.

But whatever your take on things, the set-up really is rather bizarre. For example, you'll look at a map of the northeast Lani district of the Baliem Valley and find it's divided up precisely, with the 8th Ohio Lutherans having converted and claimed the villages of the northwest, the 7th Day Adventists having the southeast, and the 34th Armenian Reformists (Minnesota Branch) taking everything in between. Every tiny, obscure religion from the most backward parts of North America has rushed out here to stick their flag in the ground and claim ownership. Just imagine the aspirant missionaries sitting obesely in their identical flat-pack houses back home, munching Burger King, wearing the sweater of a college they never went to while watching telly-evangelists on the twenty-four-hour God Channel and thinking to themselves: 'You know what, it is my dooty as a Christian to civilise those ungodly savages living naked in the jungle.'

Whether missionaries have the right to try and 'civilise' peoples who have always been self-sufficient is a question complicated by the fact that few Irianese are happy now that they have been touched by modernisation. Highlanders who have travelled to Wamena have seen the Indonesian immigrants with their satellite TVs, and want a piece of the action. Villagers once proud in their minimalist native dress now look like tramps in their grubby, shredded Western hand-me-down clothes (which house the fleas and the grime but offer little in the way of the comfort they're used to), and either cook or freeze in the new concrete-block homes they're forced to live in.

At the same time, it's hard not to admire the missionary people and their lives. Though many have managed to build themselves little Americas among the *honai* and sweet-potato fields, the life and the work here are still very hard. Nowadays, for some of the sects the conversion process is near done, and their emphasis seems to be less focused on religion and more on education and health. In the recent famines, the missionaries' airdrops of rice and medicines made the difference between hardship and starvation for some highland peoples. There's also the activity of groups such as the Croziers, who have battled against the Indonesian government in order to maintain the people's traditional art and customs. My admiration for some of these more progressive sects is slightly tarnished, though, by their preachers' habit of delivering their Christian sermons wearing native headdresses and other accoutrements. Just imagine a pasty-skinned denim-shorted pot-bellied fifty-year-old from Alabama preaching the Good News wearing a cockerel-feather crown and a penis gourd, and I think you'll appreciate the incongruity.

My own perception of the missions is coloured by my own past. I was brought up a Catholic, went to the same church every Sunday morning, and listened on the verge of dozing off to the same service right up until my teens. Weirdly, though, it was the Church itself rather than science that drove me to atheism.

One summer, my sister and I went away to a religious camp in the Lake District, to a beautiful 'Christian Bible school', with several hundred other kids from around the country. The company was wonderful. Having gone to a pretty crappy state school where the kids were generally vile to each other, I found it a breath of fresh air to be with youngsters my age who were thoughtful, kind and sincere – if occasionally a bit earnest. We played loads of sport and games, and generally had a superb time. Towards the end of the holiday I was even beginning to feel like my previous apathy towards religion had been a mistake, and that this was the thing that was really missing in my life.

But something just didn't quite click. Not one, but *every* night, these clean-cut middle-class preachers would testify, telling their tales of how they'd been down-and-out from drugs, alcohol and other evils before God had brought them back from the brink. I just didn't believe them for a second. My guess was that the closest they'd got to alcoholism was one too many sherries after Christmas dinner one year. Then they'd make the whole room stand up, and tell us all to sit down, one by one, if we felt we'd been touched by God. In a room of several hundred people, one night I found myself left standing at the end with two other kids, girls, ashamed and singled out, the girls with tears pouring down their faces at the disgrace of it. The preachers left us still standing there like the accused for the rest of the prayer class, asking everyone in the room to pray for our lost souls, and that us sinners would eventually find our way to God.

After the meeting I went back to my room, intending to cry myself to sleep with Christian guilt, but I didn't. I found that I was absolutely furious. This wasn't the 'basically be nice to everyone' kind of religion my mother had taught me. Instead it was using calculated psychology to try to entrap us. Teenagers are vulnerable, hormones popping all over the place, they are essentially awkward and unsure, but want to find themselves and be intensely passionate about something. However, they are even more desperate to belong. In other words, they are ripe for exploitation. Throughout the course of the holiday the preachers continued to ply us with emotional blackmail. Every night we would indulge in 'soaking', where everyone would stand eyes closed in ecstasy, hands held aloft, bathing in God's glory. Every day there would be heart-rending tears as nice twelve-year-olds were rent with guilt and self-loathing. It all seemed so sinister that I actually packed my bags and was all set to hitchhike home until my sister convinced me not to. It's this tendency to prey on the vulnerable that most rankles with me, and it doesn't just work with teenagers; in more recent years, someone very close to me went through a torrid and tragic time in her life, and when she was at her lowest ebb, the Church went for her like a hyena closing in on a weakened gazelle.

I returned from the holiday loathing what they'd made me feel, but determined to find out all I could about the true history and purpose of Christianity and all the other religions of the world. Over the years I've buried myself in books, lived in several kinds of Buddhist monastery, travelled to some of the world's great centres of religion and enjoyed protracted stays in the Middle East. I've danced with the harvest gods and been present at animist and other pagan rituals in many corners of the world. Though I become more and more confused the more I learn, I have at least managed to acquire some breadth of perspective on what religion is and does. I understand that gods are a human necessity and – as Voltaire stated – that 'if God did not exist, it would be necessary to invent him'. However, I still feel the bile rising whenever I see educated people trying to impose their beliefs on others.

Of course, I have another reason to wish the missionaries had not got here before me, and that's purely selfish. I travel to far-flung parts of the globe looking for uniqueness, for genuine cultures unsullied by the demons of Western civilisation. Wandering alone into a massive funeral ceremony on a remote East Indonesian island, where they sacrifice bulls, dogs and pigs in droves to their ancestors, then dance for several days in a ritualised hypnotic state – that's my idea of a travelling experience. Not slogging my way painstakingly round the world in order to turn up in a remote Papuan village and find them all reading the Good News Bible.

I'd not eaten since the orange and crackers at breakfast, and I was really starting to suffer. Thankfully the schoolteacher at Soba village managed to find me some rice and coarse salt to ease my by now agonised stomach, and one of the young lads brought me in a frog he had caught in the fields, which we boiled and shared. Its spindly bones offered up precious little meat – and watery at that – but I would have eaten it raw, I was so hungry. It was clear that they really did want to feed me, but they just didn't have enough to go round. My sense of guilt intensified still further. This was simply not how things were supposed to be. The walk out of the Baliem Gorge takes you through an area that has sustained agriculture longer than just about anywhere on the planet. It's a lush and productive food bowl not only for the peoples of the interior, but increasingly in recent years for the whole of New Guinea. In 1997, though, the widespread drought had brought devastation. And this was supposed to be the easy bit of my trip! I had mentally prepared myself for the traumas to come once I penetrated the Asmat region, but not at the stage of the expedition that should have been hardly more taxing than a pleasant stroll.

The next morning provided just as uncomfortable an awakening as before; more bedbug bites, and a stiff back from sleeping on floorboards at the schoolmaster's house. This would ease, though, as soon as I got walking, so with a good stretch, I slipped on my boots. And leapt out of them again! – they were full of huge cockroaches, which charged out over the floor and up my legs. From there on in, I've always shaken out my boots before I put them on. And for the whole of the rest of the day, every time I stopped to take something out of my rucksack a handful of roaches would scarper out and make a break for the bushes.

My first aim of the day was to hike up the walls of the gorge to get a view of the surrounding area, and out to the mountains beyond. The way up was extremely steep, with crumbly topsoil often slithering away to plant me on my hands and knees, and my hefty pack not helping my balance one bit. At one stage a dense fog swept in, slicking the ferns, lichens and mosses around me with dew, chilling the air and soaking my shirt. It was a totally different world from anything I'd seen before in the highlands: rocky landscapes with rhododendrons and tussock grasses, as well as Alpine tree ferns and wild flowers. There was also the odd-looking mountain pandanus, trees that can get to be thirty metres tall, with stilt roots and strap-like leaves sprouting from their tops. They are so treasured for their fruits that while all the trees around them are felled, the pandanus are left standing alone, their extravagant hairstyles giving them the air of a loner in a comic-book illustration.

As the fog burnt off, though, the views were absolutely worth all the effort. Although I'd not seen a Western face since arriving in Irian, I was clearly not the first foreigner to have taken this route. For upon spotting me, instead of freezing then running away screaming, the delightful children I encountered here would freeze, then break into a beautiful bold grin before demanding 'Pen, pen!' or 'One, one!' (a reference to the dollar they were hoping I'd give them for being pretty).

After another night with the pigs at Passema, I faced a difficult choice. My initial (somewhat moronically grand) plan would have seen me continuing to head downstream into the lowlands of the Asmat and on to the Korowai and Kombai regions. I had been intending to pick up information from the villages, but there weren't many people around, and those that were couldn't offer me any idea of what lay beyond the Baliem. According to my pitiful guidebook map, that way lay the tantalising badlands where cannibals lived in the trees; but at a best stab-in-the-dark guess it would probably take another three weeks of hiking to get to a decent-sized village in the Asmat.

Apart from anything else, as I'd been having to hand over substantial donations to the schoolteachers and headmen to sleep on their floorboards, there was no way my cash would last out. I was already feeling sleep-deprived, undernourished and gorged-on by the bugs, but there was no getting away from the fact that the voyage had barely started. None of the villages that I was hoping to rely on for help could even afford to feed themselves, and the further they lived from the Baliem Valley and its missionary airdrops, the worse it became. The villagers might be able to scrape together a bowl of rice or sweet potatoes, but their patience (and my cash) would wear out long before I got to my destination. So far I'd made my way purely by village-hopping, asking folk the way to the next village. But what to do when there was no next village? Would anyone lead me? There was literally nothing but jungle between me and the coast, untold hundreds of miles away. No roads, no airstrips, no cashpoints, no nightclubs, dancing girls ... no nothing. Just endless muggy malarial jungle. Run out of luck down there, and I'd be food for the flies.

Cursing myself for my pathetically appalling planning, and for naively imagining Irian would be like Indonesia as regards ease of travelling, I made the difficult choice to head back to Wamena, sort myself out, and come back with a better plan of attack.

I've never been very good at giving up. Perhaps it's daft to trace a big chunk of my character back to one childhood event, but even now, more than two decades later, the thought of it still makes me cringe.

My school was a disaster. Stuck right in the middle of the county's roughest council estate, it was staffed entirely by teachers too vacuous to do anything else, and bits of it were routinely burnt down by disgruntled pupils. I still have the scars on my hand from when I tried to grab a knife off a vile kid who was using it to try and intimidate me in the boys' toilets. I hated that school. Despite having 350 pupils in my year, because of utter apathy we couldn't even get together a full eleven players for a football team, and anything that involved competing for the school would attract as many volunteers as a certain-death mission to the sun.

I was one of my year's only sad keenos, which resulted in me at perhaps thirteen years old competing in the district sports 1500-metres event, despite never having run the race before. The whole thing came as a complete shock to me. They attacked the first four hundred metres at what seemed like a sprint, and I found myself at the back of the field puffing like an old man. I should have just kept on going and come in ten minutes after the rest – honestly, who'd have cared? Unfortunately, though, when I saw I was about to be utterly humiliated, in a crazy split-second decision I grabbed my calf as if I'd pulled the muscle. It was a stroke of genius! I limped off in dramatic agony, not having to endure the shame of coming in dead last. It was only when I saw the face of my games teacher (the only teacher I actually liked) that I realised how badly I'd messed up. And then the face of Melanie Chivers, who I'd been going out with – well, we hadn't actually gone anywhere at that point, but saying that we were going out apparently meant we were girlfriend/boyfriend. (A moot point, as it was obvious we wouldn't be going anywhere in the future either.) At school the next day I was a bit of a laughing-stock. It wasn't the teasing that got me, though – it was a dread deep shame at my own pathetic weakness. I hate being weak. Well, anyway, since the district sports debacle, I've been a bit obsessive about my fitness, and pretty fervent about never giving up on anything.

So it was, then, with sombre thoughts that I abandoned my grand Baliem mission, and headed back for Wamena.

I decided to return to Kurima on the east side of the river instead of just retracing my steps, but there was no getting over the feeling that I was chickening out. It was a good few days' hiking through stunning gorge scenery. As usual the lack of a map – or a clue – meant kidnapping a number of small children as guides, using them in relays between the villages and giving each a couple of thousand *rupiah* and nick-nacks like pens or a T-shirt for their trouble. Often they would assemble a group of friends, who would all follow chatting and laughing, the guide enjoying the new superiority afforded him by my patronage. Obviously, everyone had to have a present at the end of their leg of the relay. After paying off an entire primary school's worth of nippers on my last day, it seemed they'd got about three times what a professional guide from Wamena would have cost to guide me on the trip. Still, at least the cash was being spread about.

Later, returning into the Baliem Valley, the silence of my regular rest

4

The sun bled away from the brown river as we paddled on, the wall of vegetation lining the banks danced in the orange light, and where the sky was visible huge cumulus clouds in impossibly deistic formations grew and receded, in a palette of colours that would have made any Impressionist painter seethe with envy. Silent lightning strobe-blinked around the clouds, and swifts and martins scythed over the surface of the water snatching flighty bugs and an occasional drink from the river's surface. This was the delight at the end of a scorching day without respite from the sun and the endlessly biting flies. And then from nowhere, perhaps the most astonishing noise I have ever heard began to race up and down the banks in full-on Doppler effect. It was truly deafening, like a police siren hailing up and down the banks, remarkably synchronised and eerily piercing. Tens of thousands of male cicadas had simultaneously sprung into voice, reverberating cymbal-like organs on their abdomen in unison in a process that I may someday understand, but will never quite believe.

The Twin Otter buzzed in my ears an incessant, unrelenting mosquito whine, as the dusty plains below me finally gave way to strings of brush, then finally forest. The next airborne hour was spent looking down on nothing but lightly undulating oceans of jungle, cut through with massive, meandering brown rivers. Not so much as a glimmer of visible life. After landing, however, it was a completely different story. Senggo is the site of the largest landing field in the whole of the vast Asmat region. The village consisted of several rickety death-trap plankways teetering over dark-green ooze. There were a few wooden shacks, which could well have been knocked up in a bored afternoon, some mangy dogs scratching their ticks, and a few muscular locals moping around in threadbare T-shirts and shorts, foisted on them by missionaries years before and obviously never taken off.

It'd taken ten days to retrace my steps to Wamena, then Jayapura, and on to the dusty town of Merauke near the southeast border with Papua New Guinea. But at last I was that much closer to realising my goal of finding the remote band of clans who live in tree-houses, like the Swiss Family Robinson, more than twenty metres up in the forest canopy. The highest tree-houses are as much as twenty metres up at the tops of surprisingly spindly trees, with nausea-inducing ladders leading to their dizzy heights. Now, though, they lay upriver of me, and the most practical method of river transport was a motorised canoe. Unfortunately, alone I couldn't afford the colossal fuel prices, and anyway the freak droughts that had recently struck New Guinea meant that many rivers were impassable to engined boats. So I was stuck with paddle power, even though to my knowledge the journey had never been done this way before, so there was no intel to go on. What I knew for sure was that it would involve many days of paddling and sleeping rough in the jungle, just to get to the closest village to Korowai.

My first task, then, was to find some transport. On previous travels through Asia this had never been a real problem – with a bit of ready cash, a battered old minibus-taxi or boat can always be found from somewhere. However, Irian Jaya was already breaking all the rules of Asian travel. The people in Senggo had no farmable land on which to grow crops, and stuck to their traditional practice of gathering sago – the starchy pulp from inside the trunks of sago trees that they came across on their travels. Hence, most of the locals were out in the forests looking for some of this porridgey sludge (which tastes exactly how it looks). Those that were still in town seemed to care nothing for my hard currency. But after all, everything apart from what was found in the surrounding jungles had to be flown in all the way from Jayapura city a thousand miles away, and hence cost insane amounts – amounts only the missionaries could afford to pay. Money meant nothing to the locals, and they had more than enough hard work to do without helping some foreigner negotiate territory they didn't know, and where they would almost certainly be very unwelcome.

There was precious little here to hold my interest. Any wildlife of even vaguely edible size around Senggo inevitably found its way into the pot, but one creature that didn't was the giant golden orb-weaving

Nephila spider. The ironwood struts of the boardwalks provided a great substrate for them to make their fabulous webs. These spiders are an arachnophobe's worst nightmare, the size of your hand and with long scuttly legs. The females are monsters, sat in the middle of webs that can be the size of a table-tennis table. The threads are so strong that they can even catch bats and small birds, which they will truss up and eat just as they would an insect. Very gently, I coaxed a female off her web and on to my hands. Despite looking like they could kill you just by looking at you, they are actually very loath to bite, and if handled correctly are remarkably docile and slow-moving in the hand. In many parts of the archipelago kids collect *Nephila* webs around sticks and use them for fishing, tiny fish going for the insects caught in the threads, then getting stuck to the gooey strands.

After nearly two days of asking around for someone who'd be willing to struggle upriver with me, I was beginning to lose heart. From my initial stance of hard-bargaining tough guy, I found myself desperately offering an Indonesian monthly wage for every *day* of paddling, yet everyone I asked looked back at me with dark distrust and shook their heads. The next step would usually be for someone to suggest I go looking for 'Pak [Mr] Joseph' or 'Tuan [Sir] Peter', who might have a boat. They would inevitably point vaguely towards the other end of the village and I'd end up trudging back and forth down the walkways of Senggo, dodging the same broken planks each time and covering so much distance I might as well have walked to Korowai.

Then, after weeks hoping for rain and getting none, on my second afternoon there the skies opened, the huge drops, hammering down on the muddy, litter-filled trenches below me, seeming to mock me as I plodded the planks. Hostile faces stared out at me from the shacks and shelters. Where were the familiar friendly people inviting me inside for tea at every turn, the smiling kids dogging my heels screaming 'Hello mister!' and asking for sweets or pens? Here nobody wanted to speak to me, nobody wanted to help, and certainly nobody wanted to take me to Korowai.

It's little wonder, really, that the people here don't exactly welcome outsiders with open arms. In recent years, the outsiders who've come here have either arrived wielding Bibles or with a view to ripping the resources from their land. Historically, it was even worse. One colourful example was Italian naturalist Luigi Maria D'Albertis, who travelled up some of the biggest waterways of New Guinea in the late 1800s, living off boiled birds of paradise. He gave his name to numerous species, including the wondrous Albertis or white-lipped python, known for rolling itself into a protective ball, if threatened, with its head cocooned inside its coils. D'Albertis would drag his crew into the interior with promises of gold that never materialised. He was even better known for his total disregard for the sensitivities of the natives. On approaching a village on the river, he would fire a cannon full of fireworks from his boat. When the villagers fled to the forest in terror, he'd rampage into their homes and steal all of their most precious artefacts, even raiding their spirit houses for their heirlooms and totems.

Things got even worse when I woke on the morning of the third day and found I'd been robbed. I was dossing down on the floor of a room out back of the village store, and someone had obviously come in while I was sleeping and gone through my rucksack, stealing my emergency cash, my trusty penknife and most of the gifts I had to placate angry cannibals with. It was only the second time in over a year's travel in the region that I'd been robbed, and the fact of it affected me far more than the loss of the money. Most of my cash they didn't get – I always kept it in my money belt, usually full of *rupiah*, around my waist. The people in Asia are so friendly, and generally so honest, that I'd become very complacent about security – many nights the money belt would just have been lying under my pillow, and if that had been the case last night ... well then, I'd have been fantastically screwed. With no money at all here, what would you do? No embassy to go to, no friendly locals to take you under their wing, no phone to call in help from outside. I don't know what I'd have done.

This did nothing to boost my confidence in Senggo's readiness to be friendly. I had to get out.

The village's 'Freedom' Baptist missionaries had built themselves a beautiful if simple little sanctuary by the airstrip, with perfectly tended lawns and gardens on the coveted most elevated section of the village. They had a small bush hospital, even an ancient computer which they ran off their generator – by the look of it, it had probably once been used to land Apollo II on the moon.

I walked up the steps like a contrite sinner looking for forgiveness. Behind the fly-screens I just made out a balding American with a high, liver-spotted forehead, peering out at me like some old lady nosing at the neighbours through her net curtains. He wore his checked short-sleeved cotton shirt tucked into baby-blue slacks, with a biro in the breast pocket. He was cautiously welcoming, though he did seem suspicious I might be wanting to sell him something, as if someone would be trying to doorstop him with encyclopaedias out here in the jungle. Considering all the frustrations that were to blight my stay in Senggo, I think it's entirely possible that he had had to deal with disconsolate travellers' problems before, and probably saw us as rather a hindrance to his work.

Ken – for that was the American's name – was to be my new best friend in New Guinea. He was both missionary and medical doctor, administering to the people of the region's needs both spiritual and physical. He informed me that the biggest problems were malaria and filariasis, a disease caused by a threadworm parasite spread by mosquito and fly bites, which infects the lymph system and leads to the better-known elephantiasis. Every ghoulish schoolchild knows of this horrific disease, which produces grotesque swelling of the legs and genitals. When Ken told me that as many as 34 per cent of local people were suffering from it – epidemic level, and getting worse – I had looked at the person before me with a new respect: what on earth could drive a man apparently so fragile to come to a part of the world that was destroying *me* – a fit, young, would-be explorer – and spend decades here treating people with such gruesome infectious diseases? His conviction and calling must be powerful indeed.

Senggo and the Asmat region as a whole had made a tiny splash on the international scene that year, when Reuters carried a story round the world from this seemingly insignificant little mission station.

'It was just a few months back,' Ken began, 'and in the space of a week, we got folk from two tribes just turn up here, never heard of neither of them before, no one never even knew they existed! What do you think of that?'

I looked around at the little room with new eyes. The place had seemed incongruous to *me* here in the middle of this Jurassic forest. But imagine a hunter-gatherer, to all intents and purposes from the Stone Age, wandering out of the forest to this. That ancient computer would seem like the work of witchcraft!

'That must have been pretty scary,' I muse. 'Were you ever worried things might get physical?'

One thing that almost every explorer in the region has experienced, even in modern times, is first contacts almost always resulting in drawn bows and spears, distrust and fear leading to very frightening incidents indeed.

'No no, they were much more scared of us than we were of them. The poor guys were absolutely terrified, just kept touching everything, jumping out of their skins every time we moved too quick. I mean these people had never even seen the sea, despite living their whole lives just a few days from it. They were like kids, you know; kept putting things in their mouths to see what they tasted of, batteries, books, I dunno, everything. No wonder – well, they'd lived their whole lives *knowing* that they were the centre of the universe, that their Big Men were all-powerful and all-knowing, doing everything they could to steer clear of other tribes, and then they find all this.' Again he gestures around him to encompass his own little slice of the twentieth century in the badlands. 'They find that there's a whole world out there they never even knew about, electricity, metal, all this stuff ... a plane landed while they were here and they just ran screaming for the forest. It took them a day to come back.'

I chuckled, but then stopped myself. It seemed like a totally inappropriate reaction to other men's terror. The equivalent for a Westerner would be an alien spaceship beaming down into the middle of Leicester Square.

'We had no idea what to do with them,' Ken continued, 'couldn't speak to them other than just hand stuff.' He makes a universal gesture, hand to mouth, as if to say, 'Do you want to eat?'

'Really? How on earth do you get on round here without speaking the language?' As I ask this, I'm really thinking: Surely these missionaries don't expect the Asmat peoples to learn English?

'Oh, I speak the main local language pretty good, get by in a few dialects – and Indonesian, sure' – I feel a little guilty at my presumption – 'but these two tribes had their own languages completely, maybe spoken by no more than forty-odd people, maybe even less. It's one of the great challenges of bringing the gospel to these people – we had the whole Bible translated into the local tongue, but then you go half a day upriver and no one speaks it!'

I nodded, if not entirely in sympathy: truly, their dedication to bringing the 'savages' round to their way of thinking was impressive. I look at the water-cooler bubbling tantalisingly in the corner. Ken's demeanour doesn't encourage me to ask for a glass. Even though he's still talking to me, he hasn't offered me a seat, and I feel like an intruder.

'So what happened then? Are they still here?'

He shook his head. 'They weren't ready for the gospel yet, they need to come to terms with what they've seen first, but we have time. We still see one of the tribes every now and again. They'll come back and wander through the house, picking things up' – he gestures to the old-fashioned long-wave radio sat beside the computer. Suddenly I have a vivid mental image of a naked wide-eyed warrior pressing his ear to the speaker to listen to the static, maybe putting the handset into his mouth.

'They maybe bring a few things to trade, you know? We'll bring them to the Church in good time. The others' – he shrugs – 'they just went back into the forest. Don't reckon we'll see them again.'

All this had me stunned. A whole clan, who'd spent their lives totally isolated from the outside world. Imagine how frightened they must have been on walking into Senggo, imagine their mystification and wonder. And imagine how terrified they must have been to just leave it all, and flee into the forest. They'd be telling tales to their children for many years to come, of the hunting party that had ventured into the strange spirit world of the white man.

Outside, in the now late afternoon', the regular burps, bonks and cheeps of the post-rains frogs were starting up, and the cicadas' fierce stridulating had become a penetrating petulant banshee wail. An Asmat man in tatty shorts and polo shirt stood staring through the windows at us. Perhaps he's wearing Ken's cast-offs. Experience tells me it's just curiosity, but he appears menacing, malevolent.

'There's so much brutality here, it's such a challenge, but we've brought His word, and, you know ...' He trails off, as if I must understand. 'You know headhunting was a big problem for us when we first arrived, it was everywhere, but not so much these days. Don't get me wrong, it happens, and when it does it ain't pretty, but I kind of reckon' – he pauses as if trying to work out if he should tell me – 'sometimes it's almost just like an excuse.'

This seemed like an odd idea to me. 'An excuse?'

'Well, it's like this. Say someone has a problem with their neighbours – maybe they think they've been fishing too close to their village, or they've taken a fancy to one of their daughters. It's like the elders in the village get together, work themselves up into a frenzy, and before you know it they've gone in shooting arrows and it's got real messy. And I mean *real* messy – I mean, they'll hack a human being up so that they can wear 'em like a backpack to carry them home, and they'll just carry off the babies in a rope bag over their head. You can't believe what one of God's children will do to another, even in this day and age. It's crazy, I can tell you.'

'So is this in the areas the missionaries haven't managed to get to yet then?' I chose my words carefully here. 'Is this in areas you haven't brainwashed yet?' would've seemed insensitive, as there I was stood in his living room enjoying the cooling fan, asking him for help when he obviously had other places he'd rather be.

'Well – I'm afraid even my own parishioners lapse every now and again.' He gestured to the world outside his picket fence. I bit my lip and resisted the temptation to finish off his sentence with 'It's jungle out there.'

'It's a very dark place out there,' he continued, 'and we can't bring the light to everyone. I try and get out and see my flock as much as possible, but—' Again he gestured around him in an all-encompassing manner, looking up to the heavens in a mime that seemed to say: I have all this to take care of, and many souls to save, not to mention that fence needs whitewashing.

There was something about all this, though, that didn't tally with the symbolic idea of cannibalism that my anthropology books all told of.

'So what about the idea of taking on a powerful man's strength by taking their life and – well, kind of eating their powers?' I asked.

'No, that's just not right,' Ken replied with unexpected vociferousness. 'Most times the attacks are done at night when the neighbours are fast asleep, or they'll shoot 'em in the back with guns. It's totally cowardly, no bravery involved whatsoever. And it's more about finding new kids or wives than anything else. It's just brutality, nothing romantic about it.'

This was clearly a subject Ken felt passionately about, and not the kind of knowledge he'd merely garnered from textbooks, as I had. My next question was tentative, not wanting to be thought to have a salacious or macabre fascination with the subject: 'So have you ever had any headhunting or cannibalism round here, Ken?'

He looked at me incredulously. 'Buddy, didn't you hear about the big one?'

I shook my head.

'It was awful, one local tribe sneaked into another *kampung* at night whilst they were sleeping, and killed *everybody*. Fifteen people chopped into little bits and eaten, all the little kiddies carried off as slaves. Weird thing is, they'll be brought up as if they were their own children. The little'uns grow up knowing that their foster parents actually killed and ate their real parents, and that's totally normal.'

'So when was this?' I asked, fully expecting him to reply ten or fifteen years ago. With that, he beckoned me over to their one-room bush hospital, and showed me through the fly-screen a still, dark form lying under crisp white linen sheets. He was the sole survivor, the last of his clan.

Before I left the comfort of their little haven, cordial Ken promised he'd send one of his parishioners, who had a boat, to come and meet me. As I left, one of his young female flock brought me a picture she'd drawn of me. It showed me walking in a forest of horrors, full of snakes and spiders. I was anorexically skinny, Tipp-Ex white, and looking scared. I thanked the smiling youngster, but knew that her drawing would soon be used as tinder for an evening fire.

Ken was as good as his word, and that afternoon an unusually tall Irianese, perhaps in his late twenties, bare-chested but in what were once quite smart long trousers, came and met me at the store. He spoke some Indonesian, and we sat down among the sacks of rice out front to talk business. This almost instantly attracted an audience of maybe fifteen other men, who squatted around us and added their input at every stage of the process.

'*Selamat sore pak*,' I began. 'Good afternoon, sir.' '*Saya perbutahan naik prahu kecil kesan*.' 'I need to take a small boat to this place.'

With that, I handed him the book that had been my omnipresent tool on the journey so far and pointed to the area I wanted to get to. It was really a photoguide produced many years previously by the respected German travel journalist Kal Müller, and though a wonderful read, had never been intended as an actual guidebook. The map of the whole 4,000-square-mile Asmat area was almost featureless, just green with a few very basically delineated rivers and maybe twenty black dots marking villages. My plan had been to find a substantial aeronautical map in the country and use that. Turned out there was nothing better on offer.

My new accomplice took the precious book in his hands, the backs of which were black as soot, the palms pink as a baby's. He looked at the map for a few seconds, before turning the page to reveal a glorious photograph of an artisan working on a carving. Interest among the watchers rose dramatically, and the book was wrenched from his hands by a bearded elder, head covered in a knitted woolly tea-cosy hat, despite the heat. The crowd gathered around. Coos, aahs and tut-tut-tuts, accompanied by shakings of the head. With a ripping noise, the book was now pulled from his hands by a second man, who wanted a closer look – the pages remained with the elder. I firmly took the book back and returned to the map, this time holding it in my hands myself as I showed my potential paddler. He stared at the pages earnestly.

'I need to get upriver to Yaniruma. Do you know the place?'

This started conversations amongst the men in their own tongue, more head-shaking and pointing off in imaginary directions. Thankfully they did seem to be indicating upriver, which was a start, and the word 'Yaniruma' was being repeated. Perhaps I should elucidate, I thought.

'The Korowai people, they live in houses way up in the trees.' I mimed a house shape and pointed way up in the tree-tops. Many aahs of recognition, and repetition of my hand movements that seemed to signal understanding.

'Very long way this, many days,' ventured my prospective boatman. 'Need big Johnson.'

I was a bit taken aback at the thought that this man might be questioning whether I had a big enough penis to undertake such a journey, but then I guess that would have been no different from one of my contemporaries saying you'd need big balls to stand at the Kop end at Anfield wearing a United shirt.

'Big Johnson hundred horsepower,' he continued. *Aah.* So what he was actually saying was that we needed to take a motorboat with a hundred-horsepower engine, made by a company called Johnson. Understood.

'Two million *rupiah.*' He had so clearly just plucked this ludicrous figure from the air that I didn't even react to it. It equated to £550, which would feed an entire village for months. 'No motorboat, we must use paddling boat,' I countered, standing up and mimicking paddling a traditional dugout. This caused general hilarity, then serious faces. *'Jauh, jauh sekali.'* 'It's very very far.' Agreement all round. The bearded elder waved off into the distance, seeming to invoke a place so far that it was almost imaginary.

'It's OK.' I tried to placate them, to show that I wasn't their average over-cashed-up missionary or oilman but knew what I was doing. 'We have much time, two weeks, three weeks, no problem. I'm a poor man, I don't have 2 million *rupiah*, paddling much better.'

This last comment, translated from Indonesian back into local dialect, elicited derision and disbelief from the small crowd.

'You are a very rich man, you come here from *Belanda*' – 'Holland', general shorthand for 'foreign' – 'with many stuffs, we are very poor men.' This I couldn't really argue with. Despite us Western travellers liking to pretend to ourselves that we are wandering gypsies, taking great pride in our ability to live to a minuscule budget, we are not actually poor. The fact that we can come here at all makes us indescribably wealthy by the standards of New Guinea. My flight out here alone was more than a year's wages for any of these men. I could have argued about this being my job, but sometimes you just have to hold up your hands and say, 'Fair enough – I'm loaded.'

'So how far is it from here to Yaniruma? How many days?'

Much discussion.

'Two days,' he ventured. Vociferous denials all around, the bearded gentleman angrily raising the back of his hand as if to strike him for his stupidity. 'Twenty days,' he tries, looking at the assembled faces for approval. They seem more satisfied with his performance this time. This is getting silly now.

'Look here,' I say, putting the map in front of his face again, and staking out the distance using the width of my thumb as a twenty-kilometre measure. 'That must be sixty kilometres – say we can do fifteen kilometres a day, that should take us about four or five days.'

'Yes,' he says, brightening, 'five days.'

'And do you know the way to get there from here? Is it an easy route?'

'Yes, yes, very easy, I go there many many times.' More waving off into the distance in the general direction. Bearded fellow nods seriously and he too waves off, wiggling his head from side to side in an almost Hindu gesture of general complicity. This gives me no comfort, knowing from hard experience on the Indian subcontinent that even in India the head wiggle can mean 'Absolutely yes', 'Indisputably no', and every single thing in between including 'I have understood literally nothing you have just said to me.'

But now we were getting somewhere, it was time to talk money.

'And how much for you to take me to Yaniruma?'

Conflab amongst everyone gathered, then thoughtful silence, then someone else comes up with a salient point. And so on. Throughout it all, my slender would-be-guide sits silent and miserable, torn by forces beyond his control.

They come to a decision.

'Four million *rupiah*,' he states, suddenly exuding confidence.

'FOUR MILLION! You just said two million to do it in a motorboat! That's madness! I can't afford that.'

'*Jauh, jauh sekali*,' they remind me, and 'Much better if you had a big Johnson', as if I wasn't in possession of these facts already.

This was not going at all well. To make matters worse, my would-be-guide's demeanour was far from confidence-inspiring. He didn't seem dishonest to me, just a bit pathetic, unable to make a decision for himself. I was absolutely certain he had never been upriver before, and didn't like the idea of pitching myself out into the wilderness with him. Still, he was all that was on offer. I bit my lip and decided to persevere.

'I can give you' – sharp intake of breath – 'three hundred thousand *rupiah* to get me there and back. I'll provide all the food, you just need to bring the boat.'

It was a shot in the dark, but I knew for an absolute fact that I was offering perhaps three months' wages. Surely he would leap at the offer. Instead, he looked around questioningly at the others. They didn't like my offer at all, and were busy telling him so.

In retrospect, perhaps the reason they were so unimpressed is down to the conglomerative nature of Papuan culture. Think how weird it would be if you were trying to broker an arrangement with your plumber, and half his extended family came along too, to tell them what he should be charging. This was effectively the situation I found myself in now. It seems weird from a Western perspective, but New Guinean societies aren't arranged like ours. Their basic unit is the clan: family groups linked by birth and marriage that live together, do their utmost to be self-sufficient, and keep themselves from all contact with neighbours that may well want to eat them. Within those clans, personal property is an alien concept: everything is shared. This can sit kind of uncomfortably with our Western attachment to possessions. Time and again while travelling through New Guinea a new acquaintance would quite overtly help themselves to something of mine. Trying to get it back would elicit reactions not of guilt or shame, but disgruntled grizzles or bemusement. 'What do you mean, it's yours and I can't have it?!' But this doesn't add up to them being commune- or kibbutz-minded. Far from it.

The shared-possessions system is entirely laudable in some contexts: it works wonderfully in small self-sufficient clans, for instance. But it actually results in slightly larger settlements like Senggo seeming to exist at a lackadaisical standstill. After all, why would anyone bother working their guts out paddling a dugout canoe miles upriver in such heat, when they're only going to have to share the proceeds with the in-laws who'll just be sat there chewing betel nut?

It took an hour of bargaining to come to a price. Normally I consider myself a pretty hard bargainer, expert at walking away nonchalantly, then getting called back with a better price – and certainly never willing to settle for a price that's more than I want to pay. But none of that works when the person you're bargaining with isn't really interested, and you're totally desperate. In the end I agreed on half a million *rupiah*, which was half the money I had with me and way more than I could afford to pay. I started to think too much about what would happen if my money ran out.

The next morning I packed my bags, scowled at the owners of the store (who were probably even now figuring how much to sell my penknife for), and wandered down the creaking boardwalk to the dock to wait for my paddler.

And waited . . . And waited.

It was near lunchtime. I had sat there looking into the murky brown river all morning, occasionally taking a dip with some of the local lads, before I decided to get up and go find my guide. I wandered around Senggo, but he was nowhere to be seen. The people I asked looked at me curiously, then made some comment to their fellows, which would provoke a good deal of cackling. I didn't understand them, but the Babel fish in my ear was telling me what they were saying: 'Can you believe it, this skinny white kid actually thought he was going to take him! What a loser!' Bugger. At least grateful not to have given him an advance, I set about the whole tedious process all over again, knocking on doors, pointing to canoes and asking who owned them, even waving money in the air to try and raise some interest. Nothing doing.

That night I took myself back to the storeroom at the shop with my tail between my legs, trying to stave off the impression that the owner was sneering at me, laughing at my predicament. I locked myself into the horrid little room, rolled my sleeping bag out on the floor, counted my money and sorted it into little bundles by denomination, which I then recorded meticulously in my notebook. I neatly repacked all my gear, then ate a cold dinner of sardines in tomato sauce straight from the can. I worked out exactly how many days I could stay like this, at different levels of expenditure and with exact amounts put aside for food, fuel, telephone calls(!) and emergencies.

However you sliced it, though, I was doing anything to avoid going outside to the staring faces and the shame; I just wanted to hide away and pretend there was nothing wrong. Barking dogs and a busy mind kept me awake all night; closed eyes no more than a pretence, sleep a mere fantasy. The boatman's words rang in my mind: '*Jauh, jauh sekali* ... better if you had a big Johnson.'

5

The paddles dip and drip, trees lean over their own perfect reflections in the mirror-river, occasional swallows and terns swoop down to drink from the still surface as we glide so slowly onwards. Upriver a cappella chanting drifts towards us, and for a minute our concentration is broken. Four long black boats appear from around a bend in the river ahead. In each, silhouetted figures stand, bearing long-shafted paddles with teardrop blades, driving their war canoes on in perfect unison. As they approach, we see they are near-naked except for daubed white and ochre body paints, their muscular physiques seemingly carved from the same ironwood as their leaping war boats. This is a party with aggressive intent, perhaps beginning a mighty hunt, out to appease the spirits by the spilling of blood or seeking retribution for some untold wrong. Every stroke, every chant, the sheen of sweat on biceps and the flash of fury in their eyes, all tell of some unearthly violence about to be unleashed . . .

The next day was piercingly sunny and unbearably, skull-bakingly hot. I sat at the riverside on the flimsy wooden pier close to tears I dared not show; nearby, a few of the local kids with gurgling runny noses, unashamedly staring at me. Occasionally one would pluck up the courage to come and tweak some of the blond hairs from my arm, and run away squealing with their prize. The jetty was a stark indication of quite how low the river was currently running, towering maybe twelve metres above the slack brown stream but obviously built so that freight on quite sizeable boats could be loaded straight from deck to dock. Mid-morning, and my heart leapt as a dugout, paddled by a fit-looking middle-aged man, drew up to the muddy bank below the dock and greeted me with a humble nod and a rather limp handshake.

Here was my chance! The Senggo vultures had not yet had time to gather. Maybe I could convince the man to make this mad voyage into the unknown. This time I wasn't going to allow my prospective companion an opportunity to rethink, or even to think at all. I simply ram-raided him, proffering handfuls of *rupiah* and endless placatory platitudes, while looking hurriedly over my shoulder for the arrival of the mafia.

After briefly grilling him on the route and taking a cursory look at my guidebook map, we packed my possessions into the waterlogged bottom of the dugout. With panic, I realised that dark faces had started to gather above us. Almost immediately, harsh voices were yelling down from the dock. My paddler turned from befuddled but friendly, to frightened. I'm not sure what the mafia were saying, but it was clearly something angry, threatening in some way. Now he was shaking his head, he couldn't take me.

No way was this happening! I took up a paddle and tried to get us started. Shouts from above – they were coming down to the bank – one man reached out and tried to take a hold of the side of the dugout. I aimed a swipe at his hand with the paddle, and bellowed at him angrily in my own language. We wobbled, almost lost balance. Nearly pitching us all into the mud, I mimed to my paddler that we needed to pull, that that was the only way to stop us from toppling. His eyes were blank and afraid, but he took his paddle and then so did his boy, and we were moving, away from our tormentors, heading off towards the main river and God knew what.

A few canoe hours from the grass airstrip, the squalor and ugliness of Senggo village were soon forgotten. Nervous crocodiles were ducking beneath the surface at our approach, streaks of colour in the forest telling of fabulous birds, and reptiles hung from mossy boughs. Before us in the river, a skinny periscope with a beak popped out of the water to survey the intruders: a snake! Well, no, a body and then wings emerged from beneath the water's surface, and the 'snake-necked bird', also known as the anhinga, or darter, climbed up on to a log. These incredible cormorant-like creatures allow their feathers to become waterlogged so they can swim around submerged with just their neck and head poking up from beneath. This allows them to hunt under water without all the air bubbles in the feathers encumbering them, spearing fish with their harpoon beak. Of course, sodden feathers are no use for flying, so you're most likely to see these birds stood on riverside branches with their wings extended, drying out in the sun.

The Asmat canoe is a hacked-out tree trunk, its carcass hardened by fire. It's no more than two feet wide, and has no stabilisers. As if they weren't unstable enough, they are propelled standing upright, with a single paddle whose shaft is twice as high as a man. If women are allowed to take part in powering the vessel, they sit inside using a much shorter paddle, while the man takes the position at the rear. As he drives the boat, he also uses his blade to rudder it, keeping it travelling in a straight line.

My companion was to take this privileged position in the boat, with his son at the front and me in the middle. I consider myself quite an experienced and capable kayaker, so I was certainly not going to suffer the indignity of doing 'girl paddling', but this new standing technique was much harder than it looked. First time I got into the dugout from the bank, I wobbled dramatically, then stepped straight over the other side up to my waist in the river. My journey was almost over by the time I had even semi-mastered the technique, and for the most part I'm sure I was more of a hindrance than a help to my colleagues. While the canoe never capsized completely, I did pitch myself into the mud a couple of times, mainly after over-enthusiastic strokes that completely missed the water.

Halfway through our first day's paddling, I asked my guide what his name was. 'Michael,' he responded – he was clearly, I thought, from stock that had been visited by some of the first missionaries to the region. In fact, probably, to go by his apparent age, he may have changed his name later in life rather than having been christened as a child. First contact in this region was in some cases only within the last two decades. In order to support my hypothesis, I asked Michael how old he was. After considerable deliberation he said, '*Tidak tahu tuan.*' 'I don't know, sir.' I suggested that he was in his late thirties, and that his son Aguus was probably thirteen or fourteen, and he nodded. He was a man of few words. Michael spoke only a very little Indonesian, and seemed like a man whose mind was eternally elsewhere.

That first day we paddled perhaps ten kilometres up the main river. There was a discernable flow to it, and it looked very much like the curdled chocolate river flowing through Willy Wonka's chocolate factory. However, if we hugged the reeds at the riverside it was barely noticeable that we were struggling upstream. Occasionally, an eddy recirculating back up along the bank would forcibly carry us upstream, offering a little pleasant impetus.

There was very little traffic to speak of on the river. About halfway through this first day a longboat raged past us. The boat itself was just a dugout tree trunk, maybe eight metres long, and much deeper and broader than our paddleboat. Inside sat two very grumpy-looking women with their hair cropped close to their heads and their drooping breasts lying flat down to their stomachs under dirty grey T-shirts. In between them, a skinny dingo-like dog stood up and barked furiously at us as we passed. At the back, the young baseball-capped driver was manipulating a sputtering, roaring outboard with the maker's name, 'Johnson', written on a red flash down the side of the black engine housing. Their wake sent us bobbing up and down like a cork, and I could see them staring after us until they were just a speck in the distance. No doubt their vessel was a darn sight quicker than ours, but ours was a so much finer way to travel, under the shade of the riverside canopy, silent except for the dipping of the paddles and the screaming calls of the cockatoos as they rallied overhead.

If a lifetime of travelling has taught me anything, it's that the greatest luxury is to travel light. From childhood nights dragging the family suitcases through deserted Turkish bazaars, to adult crossings of deserts and mountain ranges, it's become crystal-clear that no luxury item offers enough joy to make it worth carrying. Go light, move fast is a mantra I've learnt from alpinism. Live simply, covet and be attached to nothing material is something I've pinched from Buddhism – though admittedly that goes right out of the window as soon as I get home and lust after the latest carbon-fibre bicycle!

My preparations for the trip had really been a question of downsizing as much as possible. I had brought thirty cans of sardines in tomato sauce all the way from Jayapura, all hidden in my hand luggage to avoid paying any excess baggage fees on the small flights. I must have made a right picture, struggling through these tiny airports with a small bag slung nonchalantly over my shoulder that actually weighed about as much as a child. 'What this bag? Heavy?' – 'No! Maybe a book or two – it's light as a feather, honest [GROAN]'!

Carbohydrate came in the form of a big box of Maggi Mee super-noodles, and the final purchase of a ten-kilo sack of rice was made in Senggo. This jungle stew has been my constant companion since my very first expeditions, and changes barely at all. You just need one pot, boil up the rice, chuck in the noodles with their essential MSG-laden flavouring packet at the last minute, and then half a can of sardines. Occasionally on this trip we would spice it up with a boiled frog, or one of the few eggs I'd managed to keep whole. On one occasion Michael succeeded in grabbing a small tree snake from a bough, and that too went into the evening's pot, fishlike bones and all. It neither noticeably improved nor spoiled the brew, which we had twice a day for the entire trip. Of luxury foodstuffs I had none at all, figuring that the enjoyment from them would be transitory but the longing thereafter would be tortuous.

As far as the equipment I was bringing with me, too, less was going to be more. Previous trips in the jungle had shown me that I didn't get on well with the recommended military jungle boots: they clogged up with mud, turning into huge concrete wellies. Once they are wet – which is instantly – they never dry out, and are a guaranteed source of trench foot and blisters. Instead, I follow the natives and go barefoot, which allows my feet to dry out as well as permitting me as much agility as possible, particularly when balancing in a wobbly dugout. On longer walks and in the evenings, I put on a pair of fell-running trainers, light as slippers without a scrap of cushioning, but with hefty grips to stop you slipping around in the squelch of the forest floor. I punched two holes in the insteps so that all the water that flowed into them would just dribble straight out again. Everything else is about simplicity: clothing and pack bought for just a few quid from army surplus stores, and with every added pocket, tag and loop ruthlessly cut off. The last thing you want is anything hanging off you to snag on the wait-a-while and hook-you-up bushes.

Routine is all-important: you have one change of clothes, your 'wets' for use by day, and your 'drys' to change into at night – a rule that must never, ever, be broken, no matter how unpleasant the outcome. Just before getting into the sleeping bag (which must always be kept wrapped in a bin liner), you clean and dry your feet, then cover them in talc, especially between the toes where athlete's foot and then the – far more evil – trench foot first begin.

I had a compass (no more than a novelty, of course, without a decent map), a cheap local-brand machete that was pretty hopeless, as I had to keep banging the blade back into its wooden handle and often inadvertently sent the blade flying clear off into the forest when chopping – luckily no one was ever standing in front of me. Then a fishing line and hooks (never caught anything), a plastic water bottle (anyone who swears by 'bladder' water systems has never had one pop in the middle of the desert), and a cheap plastic stick torch (back then, head torches were the preserve of pot-holers). My camera was an old manual Canon shooting slide film, a real workhorse that never let me down and had survived a total dousing in saltwater, recovering completely. The shutter button made a reassuring click as it engaged, and it made me feel like a foreign correspondent whenever I wore it dangling round my neck.

Shelter was a piece of blue plastic tarpaulin the size of a king-sized duvet. To make it into a tent of sorts, I'd roll a small stone in a bunched-up fist of tarp at each of the four corners, then tie a piece of string round the resultant nodule. I strung it up over a centre pole cut freshly from a mercifully straight sapling, and lie on the ground under it on a bed of cut vegetation. A mosquito net was essential, but unfortunately the local Army and Navy had only stocked them in a bell shape, designed to be used in a hotel room and hung from one point high above your head. To make it work at all under my tarp was what army boys would term 'an embuggerance', involving tying it from many places with threads garnered from palms and vines. Whatever way I constructed it, some part of me would be stuck out in the open, or pressed against the net itself, which rendered it worthless.

Last and most vital is the med kit, the one item I invest some proper cash in, and keep stowed away at the bottom of my pack. The usual workaday plasters and bandages, antihistamine and painkillers, sleeping tablets bought illegally from willing Thai pharmacists, iodine to purify water and clean out wounds; plus scalpel blades, which would be even more vital now my penknife had been snaffled. And perhaps most important, Super Glue. This wonder discovery was apparently used during the Vietnam war to suture wounds quickly and easily, and still fulfils the purpose perfectly. It can stick bits of clothing and pack back together, cover up blisters on unplasterable hands and toes, seal quite major wounds, even stick all your fingers to your face if you're not careful.

At late dusk as I paddled upstream with my muscular oar-fellows, that most incredible noise, described earlier, began to soar around the arena created by the two walls of trees lining the river. The unsettling, high-pitched scream wailed past me along the riverbank in a Mexican wave, before haring back past again and into the distance like some ethereal police siren. The volume was deafening – if we'd wanted to have a conversation we'd have had to scream. This indescribable miracle of sound from millions of male cicadas in the trees is set off by dusk's fall. It is the loudest noise made by any invertebrate, and can exceed 110 decibels – about the same as an industrial bandsaw. If you were working with this kind of noise, you would by law have to wear ear-protectors, or face irreversible damage. On my return to London, I contacted an entomologist at the Natural History Museum who told me that the performance delivered by these 'six o'clock cicadas' is so precisely timed that scientists recording the phenomenon can set up their equipment mere minutes before they know it will begin.

This otherworldly noise is most remarkable when you actually get one of these cicadas in your hand. The abdomen is pretty much hollow, and is used to amplify the sound created by the tympanums, the two cymbal-shaped organs where the armpits would be (if an insect had armpits). The tympanums are compressed by strong muscles, buckling one way then the other as many as four thousand times a minute. It's almost like Rolf Harris's wobbleboard, but just the size and shape of a sequin. The cicadas themselves are quite attractive insects, like slightly trodden-on grasshoppers, with sphinx-like pose, big spherical eyes, and their leaded-window wings fold neatly as a tent over their concertina abdomens. Some species can be as big as your hand, and if they bumble into your face, you certainly know about it! Others have glorious lichen camouflage that mirrors the tree bark they spend most of their time clinging to – the camo even continues over their eyeballs.

That evening I found on the ground a cicada that had had its abdomen eaten out by some kind of parasite. It looked as if an ich-neumon wasp had laid an egg on it. Ichneumonids are some of the most grotesque yet fascinating of all insects, enjoying intimate and intricate interactions with their host insects. Often one kind of wasp will parasitise not only one kind of insect like moth caterpillars or orb-web-weaving spiders, but a specific species. Sometimes the wasp will incapacitate the host with a paralysing sting, as in the case of the huge pepsis wasp, which in one of the great battles of nature stings giant tarantula spiders, then drags them comatose to their burrows. (But often the host will be revoltingly conscious and active throughout the whole process.) Next, the wasp lays one egg, or several, in or on the host insect, then the eggs hatch out into tiny maggots which proceed to eat their host from the inside out. One of the grimmest and most remarkable features of these maggots is that they instinctively avoid the essential internal organs of the host, in order that it may stay alive and thus keep the meat fresh.

Sometimes it seems that nature is repulsively cruel, but of course you cannot impose human sensitivities on to animal instinct. These are evolved behaviours that regulate the survival and development of both host and parasite. In fact, many scientists suggest that the presence of parasites has been the primary driver of evolution. Certainly, natural and sexual selection favours organisms that mutate or mate in a way that promotes success by outmanoeuvring parasitic attentions.

Put simply, if a cicada is born with a genetic mutation that makes it more distasteful or less visible to ichneumonid wasps, then however slight that improvement is, it is likely to help it to survive to maturity, mate and pass on its genes. Over many generations within a population of cicadas such genes will be preserved, and the creatures will move a step forward in the arms race against the wasps. Of course, the wasps are at the same time working (or rather, their genes are working) on developing more impressive stings and eyesight in order to better close the gap. If a particular population develops in isolation – say they're stuck on an island like Darwin's finches or sealed off by mountain ranges, huge rivers and so on, thus becoming so different from other populations that they cannot interbreed, then bingo! – you have a new species.

Of course it's rather more complex than this, but you can't expect the complete works of Darwin, Dawkins et al. in a couple of paragraphs!

As I examined my cicada under the tiny magnifying glass I always carry with me, I noticed in horror the cicada's head start to move and its legs convulsing, clearly (it seemed to me) in agony. 'Holy crap!' I stammered. 'It's still alive!' I was just poised to drop it under foot and put it out of its misery when a repulsive mucus-covered head broke through between the cicada's eyes, and the maggot that had eaten it alive from the inside wriggled out of its head, dropped to the floor and proceeded to bury itself beneath the dead leaves as I watched in sickening close-up.

Sometimes nature is so much more macabre than a horror movie could ever be. While these bugs offer some of the most unexpected and fascinating sights in the jungle, some of their less decorative cousins contrive to make it one of the most potentially wretched places imaginable.

These brutal riparian swamps are also the breeding grounds for many kinds of mosquitoes, and in all my years travelling through tropical forests I've never seen their equal. Mossies have four distinct stages of development. The adult females lay their eggs in (ideally still, verging on stagnant) water. The larvae develop at the water's surface for a couple of weeks, their breathing tubes sticking out of their bottoms, before emerging as winged adults that can live for several months. The crucial part of the equation is that adult females need all the nutritiousness of a blood meal before they can develop their eggs. While the males go quietly about their business of mating, feeding delicately on plant nectar, it's the females that are the vicious ones. Their evil piercing mouthparts go right through your skin – as I was finding out – and even through clothes. In many parts of the world (Scotland with its midges springs to mind!) this is just an unpleasant annoyance, but southern New Guinea is one of the most fiercely malarial regions in the whole world.

During the daytime, when the Aedes mosquitoes – which carry merely dengue and yellow fever – were at work but most of the real nasties were resting, I followed Michael's example and went bare-chested. I did, though, splosh myself head to toe with an insect repellant that was so strong it melted my plastic watch strap into a tacky mess and turned my skin burning red. The spray sweated off in minutes and would occasionally run into my eyes, giving me an itchy red conjunctivitis stare. Apparently, the olfactory receptors in our brains are the most closely connected to our memories, and this would certainly seem to be true – the scent of dry hay takes me instantly back to my childhood and the small farm I grew up on, and the smell of orange oil to my first girlfriend, who used to scent her bedroom with aromatherapy burners.

There are so many wondrous scents along the Siretsj River: fruiting nipa palms, sickly sweet figs fermenting under their mighty trees, the unmistakable heavy smell that precedes a tropical storm and the even denser humid cloy that follows one. However, the smell that always takes me back to the jungle is DEET mosquito spray. I have rucksacks up in my loft that haven't seen action for a decade, but merely open one up and that chemical smell wafts straight out, and I'm once again back on that river.

As the sun started to recede, the squadrons would awake, and from there on in it was operation cover-up. I cut a chunk out of my mosquito net and wore it over my head like a beekeeper, tucked my trousers into my hastily donned socks, put on my long-sleeved shirt, and *still* got eaten alive. The monsters would bite contemptuously clean through any material I attempted to put between myself and the environment, and their droning around my ears rendered sleep impossible and often had me swiping at the air in frantic psychotic mode. Meanwhile, my companions would be sitting almost naked by the fire, and at night would sleep outside without even a sheet to protect them.

Perhaps the most important rule is the most simple: when you get bitten, on no account scratch. I've seen so many people come back from expeditions in the tropical rainforest looking as if they have leprosy, covered with bleeding, suppurating sores that will turn into pockmarks that last for years. These were once mossie bites they've just not been able to avoid deliriously scratching. I even had one travelling companion who had to be helicoptered out of the jungles of Colombia, as the bites he'd been scraping away at first got infected, then turned into tropical ulcers. It looked as if someone had taken a power drill to his legs – you could have inserted a pencil a good inch into the holes.

So you just have to sit there with the beasties droning in your ears. They may be carrying the most deadly disease on the globe, biting you all over through your clothing. You're itching like absolute fury but unable to seek the solace of a good scratch. At times like this, the jungle is just not fun. In fact, it's torture.

At night, I slept under my mosquito net fully clothed, which did manage to keep off most of the bugs but also felt like trying to sleep in a sauna dressed in a skiing outfit. I was furiously hot and sweaty, with the bitter taste of smoke in the back of my throat and a growling hunger gnawing at my belly. As the strain of malaria found in this part of the world is the oft-fatal cerebral malaria, and has mutated to become immune to most antimalarial prophylaxis, I was taking Lariam. This evil drug seems to affect everyone in different ways – in some it produces total kidney failure. For me, though, the effects are limited to an inability to sleep properly, punctuated by spells of dozing haunted by the most vile nightmares the human mind can conjur. The Marquis de Sade couldn't come up with stuff this disturbing. Far worse than cannibals enacting their grotesque torments on *me*, I myself am the killer, chopping up children while their screaming mothers pray for mercy. With a stifled scream, heart thumping in my ears, I come to just as, beyond my shelter, cannibals with drawn bows and entrails spilling from their lips slink into the shadows.

As a child I used to make shelters in the woods out of branches and bracken and sleep out in the dark. While intellectually I knew there were no escaped serial killers from nearby Broadmoor come to tear me apart, that fear would always grip me in the dead of night, regardless. My adult reasoning mind knows the forest means me no harm, but this doesn't stop the horror tormenting me. I draw myself into the foetal position, sweaty, struggling to breathe, the wannabe explorer reduced almost to tears by imaginary ghouls and goblins.

This apart, the first few days of paddling were enchanting. Daytimes were punishing in their unrelenting scorch, rains few and ineffectual, but later in the afternoons the entire character of the place would change. As the sun threatened to dip below the trees, terns and fork-tail swifts banked and weaved outrageously over the surface of the waters, snatching floaty mayflies and caddis flies on the wing and occasionally dipping their beaks into the water for a drink, or perhaps an unseen meal from near the surface. The sun shone unconvincingly

Top left Asmat man with hat made from cuscus fur and cowrie shells.

Top right & above Bushmeat on sale in Wamena market. There was everything from cuscus to long-beaked echidna.

THIS PAGE

Above Kids playing upriver of Senggo.

Right Woman in the Baliem Valley with a piglet inside a Bilum bag woven from tree bark.

FACING PAGE

Top & centre Typical Baliem village, with honai beehive huts.

Below Carvers working on Bisj totem poles at the riverside Asmat settlement. The poles are imbued with the spirit of the deceased, then thrown into the forest to rot and return that spirit to nature.

THIS PAGE

Above Sunset over the Siretsj River, as seen from my dugout.

Left Tottering rope bridge in the Baliem Gorge.

FACING PAGE

Top Small riverside hut. Most people in the Asmat live semi-nomadic lives and stay at huts like this when on fishing or hunting trips.

Centre For one moment managing to stay upright whilst paddling the dugout – quite an achievement as I got Michael to take this photo and he'd never seen a camera before!

Below Michael paddling through a typically narrow and claustrophobic waterway off the main river.

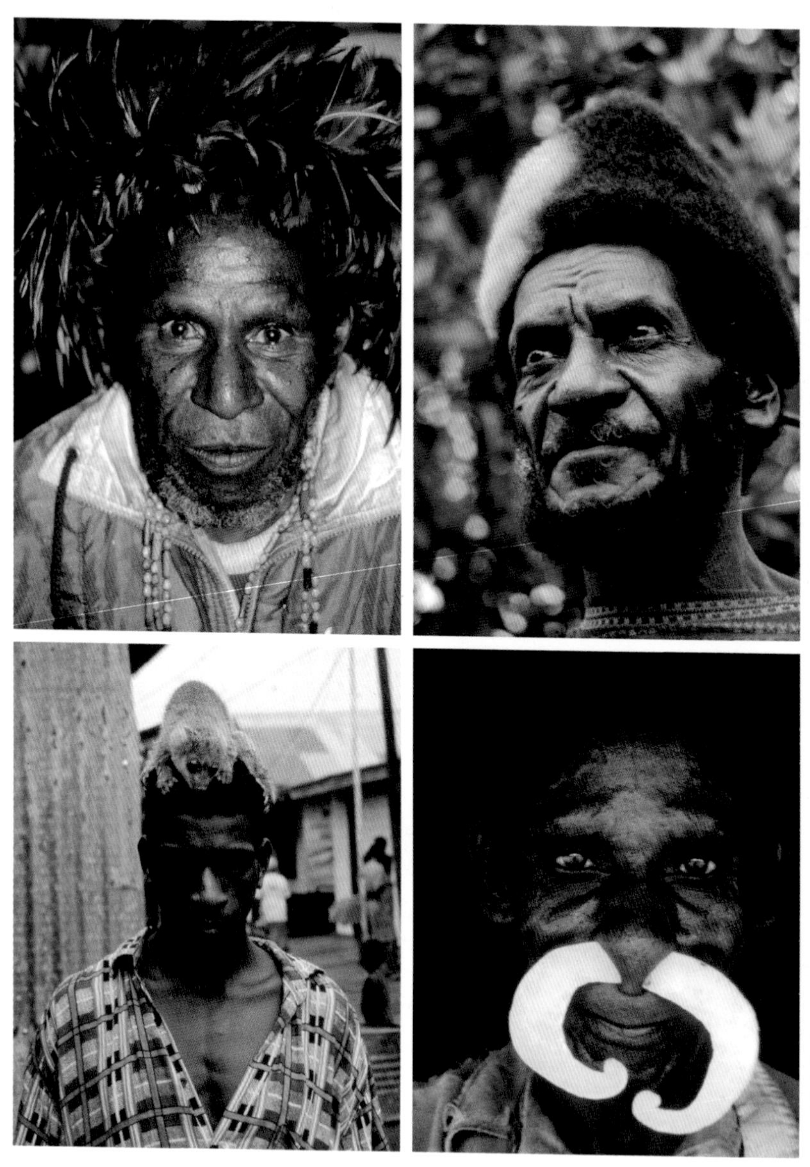

Clockwise from top left

Dani man in the Baliem Valley with a headdress of cockerel feathers.

Most of the wildlife I saw on my first trip was dead and incorporated into native dress.

Asmat man with a carved section of nautilus shell through his septum.

This man in Senggo village wandered around all day with this baby cuscus on his head or inside his shirt. In all probability it was a well-developed pouch young when its mother was caught and he decided to keep it as a pet.

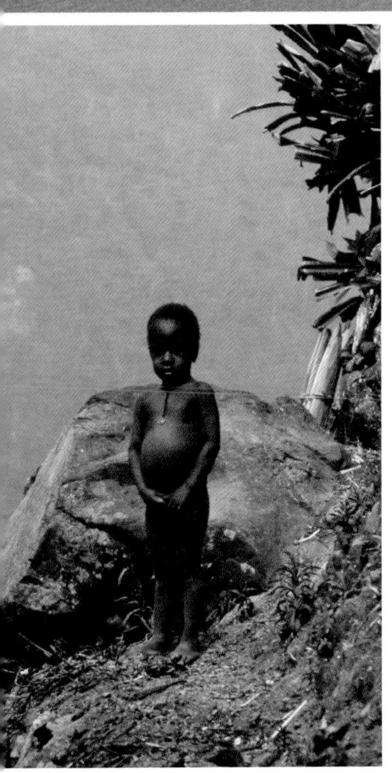

Top Sago searchers heading upstream in their fire-hardened dugout canoes.

Left Transfixed and terrified child on the sweet potato slopes that line the Baliem Gorge.

Above Cheerful Asmat children outside the smoky longhouse on the Siretsj River.

Below Senggo village, a seemingly charming village that seemed to be conspiring against me!

Above The Baliem Gorge is one of the most mesmerising places on earth, where the mighty Baliem River blesses fertile slopes, and the locals live lives that have barely changed in thousands of years.

through that thin, uncertain high cloud known as altostratus, leaving a perfect corona around it. The diffused, softened light reflected perfectly in our glassy river, so that it seemed the brown flow had turned watery blue, with whispers of gold and nectarine. A great billed heron stood motionless like a pterodactyl in the reedbeds, before stabbing into the murk to impale fish. He too stands over his own perfect mirror image. As he dips his bill towards the water he seems to kiss his twin, and then the surface is disturbed and the vision disappears. The piercing calls of dusk are not yet in voice – the only sounds are the excited chirrups of the feeding swifts, and occasionally, when they soar over my head, the roaring of the wind through their scything wings.

In this tableau of wonder four long silhouettes now appear, bearing down on us at a pace, seeming to skim along the surface of the river. As they draw near, figures paddling war canoes become clearly visible. The long, sleek vessels are each made from a full tree, and contain five or six standing men, drawing their long-shafted, leaf-bladed paddles in perfect unison. At the prow of each boat is an intricate carving – a highly stylised phallus, a fierce crocodile or a mighty warrior. Their chants carry far across the silent waters. Painted white across their chins, throats, foreheads, backs and shoulders, they wear skirts of grass and headbands made from yellow and spotted black cuscus fur. As they draw close and catch sight of me, the chants turn to squeals and whoops – sounds of Masai feast change to Cherokee war cry. The blood thickens and cools in my veins.

I scrabble for my camera, but they are already past, their chants fading into the distance. Why were they dressed as if for war? Why were they headed downstream with such purpose? The beating of my heart slows. I'm secretly glad they continued on their way, that they didn't find their objective here and now.

This area is named after these peoples: the Asmat. There are many possible meanings for the name, but it seems the most probable is 'tree people', from the conjunction of the word *os*, meaning tree, with *amot*, which is a compound word for 'man'. The Asmat people's affinity with the trees and the jungle is absolute. Many inland tribes live their entire lives in the tree-tops; trees are even seen as representing people, with the roots as feet and the fruit as the human head. Much as a tree must lose its fruit in order to produce a new tree, in darker times Asmat warriors would have to take the head of one of their enemies to ensure that new life was born. An enemy's brains are thought to pass on the powers of their owner to the consumer, and an ancestor's decorated skull is a protective talisman. Headhunting was, and in some remote places still is, an integral part of Asmat life, and though some tribes may wait several years between raids, a crop of enemy deaths is essential for the Asmat to maintain their equilibrium with their ancestors' and their environment's spirits.

In 1961 when Michael Clark Rockefeller went missing in the Asmat jungles, it was considered that he had fallen foul of this same ideology. A high-profile member of one of the richest and best-known families the world has ever known, Michael was the son of New York's governor general, later to become vice-president of America. Michael spent much time in Western New Guinea, on expeditions for the Harvard Peabody Museum and also collecting tribal art. The facts of his disappearance were as follows.

Rockefeller was off the southern coast when his vessel's engines were swamped and the boat rendered directionless. Two shipmates swam ashore, leaving Michael and a Dutch anthropologist to drift through the night. The next day, the shore was just visible, perhaps three miles away. Michael was a strong swimmer, and with two oil drums for buoyancy he struck out for the shore like Dreyfuss and Schneider at the end of *Jaws*. Rockefeller was never seen again. His disappearance here in this forgotten place caused international intrigue, and inevitable speculation. Had he been taken by headhunters? Maybe even consumed by cannibals who wanted a symbolic part of the fantastic power he represented? If he had made it to shore from where he was last seen, he would have arrived exhausted near the village of Otsjanep. Naval and missionary boats scoured the seas, along with a thousand Asmat canoes. Twelve aircraft and many helicopters were dispatched to do an aerial survey. Rumour has it that three skulls of Caucasian descent were eventually recovered, but Michael's body was never found and the case was never resolved, though he was declared officially dead three years after his disappearance. Inevitably, though, many years after, myths persisted of his living deep in the jungles as a sort of sinister deity, in some *Heart of Darkness* dystopian vision.

On the face of it, it should have been a simple story: man tries to swim three miles in rough current-ridden seas that are home to several potentially man-eating shark species and the largest crocodiles in the world. It seems likely to me that he simply would never have made it to shore. In his book *Ring of Fire*, however, Lorne Blair reports that just three years earlier Dutch patrol boats had sought to curtail tribal wars in the area by strafing Otsjanep with gunfire, killing amongst others four of its war leaders. Blair is convinced that the seas off Otsjanep are so shallow that Rockefeller could have practically walked ashore, and that he was killed in an eye-for-an-eye revenge on the white man.

That night as I slept in the riverside forest I was woken by the sensation that my ear had caught fire. I leapt up, thrashing at my head to put out the flames, but found nothing, and the pain was spreading down my face and shoulders. Firy stings covered my face, and as the pain subsided, a rush of adrenalin left me feeling jittery and then strangely weak. A column of red ants had crawled into my sleeping bag, and were fastening themselves to any part of me they could find, biting with snappy little jaws, then stinging straight into the wound they'd created. Bug spray had no effect. So I had to vacate my sleeping site and sit out in the dark for a few hours until they'd passed on. The next day every bite came up in a small red spot with a perfect white pus head at its centre. It was too tempting to squeeze them, and they ended up as ugly, angry welts.

The following afternoon, we pulled in to a riverside longhouse, a thin, rickety building with a porch-cum-landing jetty hanging over the murky brown water. Access was via a single wooden pole at a 45-degree angle with footholds cut into it, which was quite a challenge in itself, not to mention holding my camera and my other worldly possessions while desperately hoping not to pitch over the side into the river. It was never really apparent if my paddler and his son were actually related to, or even knew, the clan who owned this communal longhouse; their greeting was certainly not particularly familiar. Their initial reaction when they saw there was a white man in the canoe was pretty dramatic: the women covered their naked breasts with their hands and sheets woven from wood bark. Older children ran away, younger ones froze like rabbits in the headlights of a ten-tonne truck.

The longhouse was about twenty metres long, with thatched palm roof and walls and a bamboo floor covered in places with bark mats. It appeared to be home to about forty people, all living together, male and female, adults and kids.

I made the mistake of kneeling down in front of one little girl and making to shake hands with her. She erupted into horrified wailing and screaming, tears pouring down her face as I writhed in embarrassment. She just would not stop, and the more I smiled at her and made conciliatory gestures, the more she yelled. I didn't know which way to look. The men were bolder, staring at me with a mixture of hostility and mistrust, all totally naked but for their bows and arrows. It was a nervy start – particularly as the kid still wouldn't stop screaming.

We all retired inside together as the sun faded. The interior was dominated by several open fires with no chimney, which made for a thoroughly noxious night but certainly deterred the mossies. Instead of cooking up from my own stores, I decided to ask if I could buy some dinner off the inhabitants, and they willingly agreed. They returned with a plate of dried catfish heads that stank – well, like old fish. Unfortunately, having taken the meal, manners dictated that I had to eat it, so I munched gingerly on the meagre pieces of bone, face puckered like a gummy granny sucking lemons.

Dismissing worries about the giant saltwater crocodiles that prowl these waters, I had myself a good swim and washed off the crustiness of the day. On a nearby muddy bank, a group of children were sliding down into the water, and using a stunted tree as a diving board. After some initial nervousness, they started showing off to the strange white man, doing extravagant backflips and jumping off their friends' heads. Here was a way, finally, for me to get properly involved with the locals. I clambered up on to their diving log and did a poor comedy impression of their leaping and bounding, incorporating a whole bunch of pratfalls that would have made Buster Keaton proud. Within a few minutes the barriers had fallen completely, and they were using me as a climbing frame. There I was, buried in naked brown children, who were clambering up me to dive off my shoulders and trying to balance on my head, white teeth flashing in unabated laughter. It was the first genuine connection I'd made with the locals, and probably the happiest I'd been since arriving in New Guinea.

The kids were totally tireless. Once I'd picked one up and thrown him into the water, every one of them had to try, and again and again. After perhaps an hour of being diving board, climbing frame and trampoline, I was totally exhausted and forced to retreat to the longhouse to escape their boundless energy. In the firelight, I sat watching a couple of naked artists wreathed in smoke, carving extraordinary wooden sculptures of tangled limbs and cavorting figures. One of the men, with the physique of a prize-fighter, wore a carved nautilus shell through his nose, to give him the intimidating appearance of a wild boar.

The nautilus is a cephalopod, related to the octopus and squid, and has a large, beautiful multi-chambered shell, white with red-brown tiger stripes and twisted like a dinner-plate-sized snail's shell. The nautilus, though, swims deep in the water column, rarely less than eighty metres down. They can be baited for and caught, but how any of the shells get this far upstream from the sea is a conundrum. I guess it's to do with the fact that the main reason these people live right on the riverside is to gain the benefits of trade, moving upriver from the coast to remote inland communities.

The Asmat carve all sorts of wooden wonders, including drums, shields and masks – mostly from the valuable local hard ironwood – of which the most interesting is the *bisj* pole. A kind of totem pole, the *bisj* is cut in a single long piece from a buttress-rooted mangrove tree, then carved with naked human figures climbing all over each other. The flag-like buttress root at the top is crafted into an intricate representation of the phallus, jutting proudly from the uppermost human figure's groin. These poles are thought of amongst collectors as some of the finest such pieces in the world. However, they're not seen by the Asmat as being art to keep, because they are carved for the recently deceased. The local witchdoctor will perform a ceremony translocating the dead person's soul into the pole, which is then taken into the forest and cast away to rot. As it does so, the spirit will be released into the environment it has come from.

In Asmat rite-of-passage festivals, boys have to complete a number of rituals before they are considered to be men. First, they are sodomised by all the village elders, before being sent out to neighbouring villages where they are obliged to take part in widespread slaughter.

They generally take the heads of the men, kill the older women, and appropriate the children and women of childbearing age for themselves. In tiny, isolated clan-based communities these exchanges of human cargo and disparate genes are perhaps all that has saved the people from dangerous inbreeding. A successful headhunting raid would be followed by a huge feast, with orgies and wife-swapping. Then, the young headhunter marks his step into adulthood by smashing his trophy skull open with an axe, to spoon out and eat the brains. This is said to endow him with the powers of his victim, while sucking on the penis of the corpse imbues the victor with their masculine virtues.

Anthropologists have suggested a more nihilistic spiritual victory, whereby your enemy's soul is totally annihilated by being eaten, then further degraded when his remains are defaecated. Most of this is supposition, though, as evidence is mainly second-hand, dragged as it is through the distorting mirrors of language, culture and judgement. In addition, great care has to be taken when inquiring about the subject, as people are very easily offended. For many it's their filthy, shameful past, whereas for others, it's their sacred, secret present.

Particularly in Indonesia, there are many places where you'll find versions of imported religions such as Christianity that have merged with native beliefs to form new hybrids. It's not surprising, really, since it would seem that all the globe's religions have been born out of the same universal human needs – for example, to offer explanations as to how the earth was made, to provide ritual and structure to human life, to find a reason for life, to stave off the fear of death and to assuage the existential angst of feeling alone in an indifferent universe.

For all these reasons it's not always been that difficult for zealous Westerners with a Bible and a little bribery to convert animist natives to at least nominal Christianity. However, when the missionaries decide they've had enough of the jungle and head back to civilisation, often native tribes will just assimilate their new-found dogma with their ancient animist traditions. In far-flung outposts of Indonesia I've participated in harvest festivals where goblets of pig's blood are passed around the congregation as a pseudo-sacrament, funerals where hundreds of animals are sacrificed – often on stone altars – and ritualised mourning combined with elements and sometimes the language of Christianity.

Here in the Asmat region, neglected converts from a headhunting past have assimilated to some degree the ritual of transubstantiation, where bread and wine become the body and blood of Christ and are consumed. Taking the Eucharist has for the Christianised Asmat become a replacement for, and perhaps symbolic of, cannibalism, whereas some non-Christian Asmat tribes now eat sago grubs ceremonially instead of consuming an enemy's brains (which the puffy, wrinkled white larvae convincingly represent). In the highlands of New Guinea, where locals have for millennia consumed parts of their family members upon their death – to make sure their spirits and powers live on – the same metaphorical connection has been made. Some modern anthropologists suggest that the concept of communion was itself invented by early Christian scholars as a means of enticing into the Roman Empire widespread pagan communities who may have practised actual or symbolic cannibalism.

Fascination with the subject is something I share with every traveller who's ventured into these wild lands. Perhaps it's a desire to find something primal, untamed, taboo in a sterilised, homogenised world. Maybe it's because these people are the living evidence of how we used to be, and are somehow deeply connected to me through some version of the Jungian collective unconscious.

Remarkable evidence has been found in prehistoric finds over recent years. Large numbers of excavated human bones from different places have many features in common: for instance, those found from around 800,000 years ago in northern Spain, from near Cheddar Gorge dating back eleven thousand years, and near Eton in Berkshire dating back merely to the boundaries of the late Iron and Bronze Ages, three thousand years ago.

These bones show cut marks made shortly post mortem which tally with the cuts on animal bones made when processing or preparing an animal carcass for food. There are signs of defleshing, skinning and dismembering, and of the bones having been smashed apart to access the marrow. Jaw bones show cuts to remove the tongue in an exact same manner as happened to deer jawbones found in the same midden. And these middens themselves are significant, containing as they do bones not found as complete skeletons, but littered in amongst animal bones and other food remains. Scientists suggest that cannibalism has been at some stages in the human past very much the norm. Amongst our closest relatives, the chimpanzees, cannibalism has also been recorded. I've several times had the privilege of encountering chimps in the forests of central Africa, and it is the most primeval and disturbing experience imaginable. When chimps go on the hunt, their excited screams of blood lust pierce through to a place that (to quote Jack Nicholson) 'you don't like to talk about at parties'. This surely is nothing, though, to the horror of seeing an alpha male chimpanzee rip an infant from its mother's breast, smash its skull against a tree trunk and then consume it.

A fierce-faced young man was sat on the balcony stringing a hook on to his line, and I asked him if there were many fish in the river: *'Dan ada banyak ikan disini?'*

'Banyak, banya,' he affirmed with a nod – 'Many, many' – and motioned with his hands that they were about a foot or so long.

'You have to wait long before one bites?'

'Ndak!' he retorted proudly. 'No way! I wait maybe a minute, two minutes, then they bite, and we eat.'

'And you pray it's not a crocodile on the line?' I quip, miming him catching a fish, looking chuffed to bits, then suddenly terrified as the imaginary line yanks him forward and makes to drag him waterskiing across the river. I laugh at my joke, encouraging him to join me in merriment.

'Tidak ada,' he replies seriously, returning to his hook. 'Nope'. He clearly thinks I am not quite the full ticket.

Inside, I sit around the smoking fire watching one of the carvers work on what appears to be a drum of some kind, using a dark wood that looks as though it has been soaked in creosote for months. The work possesses the practised simplicity that lovers of so-called primitive art so value. The crocodiles and the people he depicts have little or no detail to them, but you could never doubt what they are, for they are perfect icons, or symbols, of the spirit embodied in what a crocodile really *is*. Try and imagine a highly trained artist imitating the brushstrokes of a child's painting: the crudeness is deliberate, expressionistic, almost disturbing.

'How long do you work on a piece like this?' I query.

'*Barangkali dua bulan*' – 'Maybe two months.'

'Two months! Wow! This is very beautiful,' I add in admiration. He nods, and continues carving.

'How much for this drum?' I ask, more out of interest than from actually wanting to buy it. God knows how I'd get any art out of the Asmat region anyway.

'Two million *rupiah*,' he replies.

I gasp.

Looking around me in the huge longhouse, I see nothing of material value whatsoever. The people are eating fish they catch themselves, with rice. Two million *rupiah* would keep everyone in the clan for a year. Does he honestly command prices like this? I know that Asmat art is internationally acclaimed, and in all probability dealers come up here regularly to pick up the work and barter or pay for them. But *two million*? For a carving? That's about £500!

'Thank you very much Tuan – it's very, very beautiful,' I say graciously, with all the fulsome praise I can muster. He nods, but doesn't look up.

I go out on to the broad balcony over the river. The sun is almost down, and the mosquitoes have started biting. The sunset stains the sky with a medley of soft fruit flavours, and I see the comet Hale Bopp painted above the horizon like a white lollipop. It's beautiful, and this is the kind of moment when I would normally be overcome with enthusiasm, with wonder at the beauty of the natural world, and start waxing lyrical about the rightness of such a simple life. But I don't feel that way. Fact is, I've spent time with various tribes all over the world, and I've usually had a strong sense of the perfectness of their union with nature, the uncluttered simplicity of their lives. I love the easy smiles, the free revelling in their pleasures and the absence of pressure from modern mores – they make me question everything we value in the so-called civilised world. But not here. In this bug-ridden hut teetering above the mud, in the choking smoky interior – the only escape from an unforgivingly scorching sun – it didn't feel like Shangri La at all. In fact it felt an altogether brutal existence, battling an environment that wanted to hurt you. Even against the glory of a New Guinean sunset the people seemed sullen; and what was more, I felt fabulously unwelcome. Every second, I was being made to question my motives. What was I doing here in this place that didn't want me? Was I here just to gawp at the natives with the bones in their noses, or did I have something idiotic to prove to myself?

Later, as we bedded down, one of the men of the house laid his head on a bleached and painted skull, its eye sockets plugged with tar and studded with beads and small seashells. It was the skull of an ancestor, which he'd chosen to use as a pillow to guard off evil spirits.

That night in the longhouse was, to say the least, bloody miserable. Not because I saw myself becoming longpig casserole – although they did turn me into a six-foot kebab. My bed was a solid wooden platform about a metre above the floor, underneath which was my own personal fire of smouldering coals. It continued smoking all night, supposedly to keep away the mossies. The fireplaces were made of mud slapped on to the wooden floor, and had baked into a (hopefully) fireproof clay. I lay there tossing and turning, every move sending an acrid blast of black smoke into my nose and lungs.

So the night had begun with me sweating like a pig on a spit, but at about 3 a.m. I woke in a cold sweat – my hyperactive nightmares had plunged a cannibal's spear into my eye. There was no relief, though, in realising that at least the spear was only a nightmare. The longhouse was pitch black, the slumbering forms about me barely visible in the meagre leak of light from the open doorway. I fought to swallow the thick air down my smoke-seared throat, the dark walls threatened to engulf me, and the thatch of the roof was suddenly a seething mass of huge and terrifying insects. All of a sudden, with the stark realisation of the horrible potential of my situation, total panic overwhelmed me. I was utterly alone, weeks away from anyone who spoke my language or might come looking for my rotting corpse. If these unfriendly individuals chose to spear me for my cash and my camera, no one would be able to trace me to within a thousand miles. I tiptoed out on to the porch, almost crying with the desperate effort of silence and the dread of waking the men inside.

Outside, the broad river swept silent and oblivious in front of me, a little cloud-filtered moonlight glinting on its depths. The jungle around me was another mass of horrid potential, its eternity of tangles and claws alive with slimy creepy-crawlies, a billion tiny blood-sucking evils. I sat on the jetty clutching my knees, fear gripping me as it had never done before.

With dawn, the sun rose and burned away the nightmares. Only the sleeplessness and the stinking grime of sweat and soot in my pores remained. That, and a hacking smoke cough that lasted three days. As I left the longhouse there were no goodbyes, no children gathered to wave us off – truth is they were at the very least ambivalent towards us, and we had perhaps even been an annoyance. So much for life-affirming experiences.

There was no getting around the fact that I was missing home. I'm not ashamed to admit it, as homesickness is not an affliction that often haunts me. This isn't because of anything lacking in my home and family – quite the opposite. I figure that the total support and stability I've had in my home life has determined my adventurous nature. Subconsciously, perhaps, I've always lived under the comfort blanket of familial security, knowing that if any of the massive gambles I've taken had failed to come off, the family home would always be there for me. It sounds so pathetic when I put it this way: 'I'm adventurous because Mummy and Daddy would always be there to come to my rescue'! But it's a totally subconscious thing. And in one sense it's totally illusory, because if anything had gone wrong out there, as on all of my expeditions I would have been utterly on my own. I missed home with a dull, miserable ache, and the additional mortification of the weakness this signified. Every time a vision of home cropped up in my head, I'd growl at myself through clenched teeth, practically beating myself about the face for being so feeble.

There was so much to miss, though. As I've said, I'd had an unusually charmed childhood, with so much happiness. Things were a struggle for most of our young lives. Mum and Dad were not well off, and we lived in a two-bedroom end-of-terrace house, with Dad having to take second jobs at a pig farm and the next-door pub just to get by. Our toys as kids were crafted and carved into being out of such items as skirting-board offcuts, yet seemed the most magical playthings we could want. We were easily pleased. An obsessive reader, I buried myself in tatty jumble-sale books; ideally, I'd choose tales of wild deeds in wild places, but would consume anything from romantic fiction to history to encyclopaedias. We wanted for nothing, had love in spades, and simple but firm rules to live by: 'Treat others as you would be treated yourself', 'It's better to give than to receive', 'There is nothing more criminal than waste, and nothing worse than wasting time', 'Better to try and fail, than to wonder what could have been' – and most importantly, 'It doesn't matter what you do in life, as long as you do it with all your heart and soul.'

When I was five years old, though, fate tossed the Backshall family a remarkable gift. We were puttering up a hill in our ancient dark-green Austin A30 (if you can't imagine it, it's like a Morris Minor without the character). The car was inherited from our great-uncle, who had had polio, so it was worked with hand shifters. It was only recently green; Great-uncle Phil had kept it in a shade of depressing grey with one odd brown door, so Dad had taken a can of dark-green emulsion and a brush and repainted it. Rather messily. It was quite something, that old car: you could see the road passing beneath through rusting holes in the floor, and in the winter when it got really cold you had to start it by cranking it with a metal handle that went in just under the front bumper.

Anyway, out of the car's windows that day we saw a sign at the side of the road: 'Manure, 10p, bring your own bags.'

We ground our way up the rough driveway, running like a tunnel through the thick rhododendron bushes, to an enclave of ramshackle buildings surrounded by towering oak and silver birch trees. We parked the car at the base of the hugest oak, which glowered over some hen coops (and unbeknownst to us was to be the car's eventual resting place when it finally gave up the ghost years later). The main farmhouse was little more than a hundred years old, though styled from Tudor times. I later learnt from a historic map of the area that the small blue cottage in front of the main house had been the area's post office for at least four hundred years. In front of the cottage itself was a long lawn and an apple tree in full and magnificent blossom. A line of damson and plum trees bordered a chicken run the size of two tennis courts and full of ducks, guinea-fowl and chickens.

Beyond were paddocks where two beautiful grey horses and a serene chestnut pranced – the source of the manure heap that we had come to plunder. An honesty box was nailed to the side of a stable block. Two peacocks were sitting on the roof, looking and sounding both incongruous and perfect at the same time. The large wooden barn had a ground floor with more stabling for the horses and the tack room. Up a steep ladder and through a trapdoor was a huge, dark hayloft with centuries-old cobwebs, ankle-deep dust, prodigious colonies of rats, ancient junk and stacks of *Horse and Hound* magazines. In later years I was to try (and fail) to lose my virginity up in that hayloft, and it was the site of endless parties, get-togethers and sleepovers – the perfect place to hide from parents' prying eyes.

Bright-blue paint along the rafters failed to cover up the fact that they and almost everything else were deeply rotten, and probably could do with being condemned. The whole place seemed like a vision, and my parents were both starry-eyed the second they saw it. As Dad filled the bags with horse dung for the roses, an elderly man with the kindest face I can recall from my childhood came down, introduced himself as 'Don', and chatted to the handsome young couple my parents would have made. My father was bearded, athletically built and always tanned, a simulacrum of the young pre-alcoholic Oliver Reed. My mother had impossibly deep-brown eyes and a smile that could charm a genocidal dictator. But Don was most taken with my sister and me; hopelessly over-enthusiastic, polite, confident kids, always glowing with pride in the presence of our parents. Don was struggling to keep the place, alone as he was in his advancing years. Worse for him, as the farm was owned by the Crown and on green-belt land, he only had it on a lease, which had seventeen years left to run. After that he would be left with nothing, and the uncertainty had been a cause of concern for many years now. Of course, I wouldn't have taken any of this in. All Jo and I knew was that when we got back into the car a few hours later, it was to the exciting news that we were to swap our tiny terraced house with Don, and move into Collingwood House.

Mum and Dad took up the management of the place more in the way of a lifestyle than just a home. They bought a thick book on self-sufficiency, planted an enormous vegetable patch, and embarked upon the impossible painting-the-Forth-Bridge task of rebuilding the dilapidated buildings. After a few months, our first rescue animal arrived, an asthmatic donkey called Barney, and thereafter the flood-gates opened. We collected all sorts: dogs, goats, floppy-eared rabbits, two intimidating geese called Victoria and Albert, and a malevolent Exmoor pony called Walnut who deliberately headed for low-hanging branches to try and forcibly remove anyone daft enough to try and ride him. My sister developed a very intense relationship with him – every time he manoeuvred around to kick her she'd boot him first, and eventually had him following her around like a soppy puppy. Me he would just try and kill at every available opportunity.

There was a sad little duckling called Twit who imprinted on my mother and would follow her everywhere, even sitting in the sink when she did the dishes. I say 'sad' because one day we inadvertently locked her out when we went to school, and came back to a few feathers. Presumably she had been fodder for a stoat, or the insatiable foxes. And this was far from our only failure; despite Dad digging the chicken fences down deep into the ground and the fences themselves being eight feet high, we lost three complete sets of chickens in guerrilla raids by foxes – we could hear the foxes screaming at night like ravished damsels. In an unspeakably macabre tableau guaranteed to provide material for years' worth of nightmares, both the rabbits hanged themselves after an ill-advised experiment with putting them on leads. A young fawn got itself tangled in the hen-coop wire and was badly injured, to the distress of its mother nearby. We took it in and desperately tried to nurse it back to health. Adorably, our dog Buster wrapped himself around the fawn to try and protect it, growling at anyone who approached too close. The fawn unfortunately didn't manage to pull through, and was buried in the sadly expanding pet cemetery of the chicken run.

Buster himself was a nightmare. We'd got him from the RSPCA centre where I was then working, and hadn't had the heart to neuter him, so this roaming Littlest Hobo would turn up every other day in the next county, having followed the scent of some lady dog or other for miles.

We nearly lost our goats several times to rhododendron poisoning, and did lose several kids to stillbirths. However, Mum did an extraordinary job of delivering several of the youngsters herself. Perhaps my starkest childhood memory was watching her as the birth of our first ever kid went dramatically wrong, the panic as Mum and Dad realised it was a breach birth, that the kid was coming out the wrong way, and would surely die unless Mum did something about it. I remember her propping up the by now dog-eared self-sufficiency handbook on the windowsill so as to follow its instructions. It was dark outside, and the light inside the tiny stable was faltering and dim. She pushed her hands right inside the straining, bleating mother, pushed the kid back in and turned it around. Then suddenly there was a plop, and a sack of wet, pink slime dropped out into the straw. It was totally lifeless. Jessica, the mother, turned and licked at it, exhausted, and finally it raised its head weakly and started to move. We cheered and cried and hugged each other with joy. Months later, Blossom – born as the apple tree bloomed – was bouncing around her meadow with the unbridled joy young creatures all seem to show in springtime.

All the animals were much more pets than they were farm animals. Even the ducks and chickens we knew individually by name and personality, and to my mother's chagrin the goats and horses found their way into the house. As the goats' enclosure didn't seem large enough for them to get enough exercise, we'd take them for regular walks up through the woods and out to the surrounding common land. This drew stunned reactions from local dog-walkers, as they saw us approaching with a big white dog, then when they got closer realised it was actually a bleating Saanen goat.

We used to play hide-and-seek with the goats, running off into the bracken when they weren't looking, then sitting quietly waiting. Just minutes later, a wet nose would gently nuzzle into your ear, as Blossom, perhaps, would come and find us. It was my job to milk the goats before going to school each day, and it was a lot harder than it looked. If you left it for even a day, their udders would be swollen and clearly painful, but far from being grateful that I was relieving them of their burden they would wait malevolently until the pail was full before giving a big kick and drenching me in warm, frothy goat's milk. If it was a warm day at school, later on I'd end up stinking of curdled milk. Not surprisingly, my nickname at school was 'manure boy'.

Our childhood wasn't all halcyon days, though. There was illness and arguments, tragedy and tough times, death and disaffection, as there will be in any family. But through it all shone the ceaseless wonder of the place: the changing seasons marked by the blossoming, fruiting and shedding of the trees, the arrival of various animal babies in the spring, the visits of the horny-handed farrier, and the festival atmosphere as we gathered friends together to toss a hundred bales up into the hay loft in late summer.

Even as a child I dearly valued my own space, and the woods were my retreat. Wondrous ancient woodlands of conifer and broadleaf, dappled sunlight, the scent of pine sap, the scurry of squirrels. The fluting of blackbirds even now takes me back instantly to throwing my windows open on spring mornings, and the blue tit who tapped insistently on my bedroom window every day for two seasons running. I knew the location of every fox earth and badger sett, stalked red and roe deer to see how close I could get, and cried when one of the stable girls set light to a clutch of grass snake eggs found in the manure heap.

For the rest of my life, no matter where I go or how much I make my current house my own, Collingwood House will always be my home.

Back paddling down the muddy river, and progress was becoming stultifyingly slow. The heat was totally overpowering for most of the day, and a dose of the runs had me feeling weak and empty of fluids and calories. Even the worst of conditions can be endured when you're feeling fit and healthy, but when you're sick, all you want to do is curl up and be taken care of. Suddenly, a flash of iridescent green feathers passed in bouncing flight in the low canopy above our canoe. Michael became wildly excited, ditching his paddle to scrabble around for his spear, rocking the dugout and threatening to plunge us all into the mud. '*Cenderawasih!*' he shouted – 'Bird of paradise!' – but it was already gone.

This was my first tantalising view of the mythical birds that as a child drew me towards Papua New Guinea above any other destination on earth. The allure of vast, dark rainforests, inhabited by jewelled birds like flying angels, was my most romantic vision. But unfortunately, they are a sight that my children will probably never have a chance to see. There are forty-two species, thirty-six of which live only here in Papua. The birds have been hunted constantly over the last nine thousand-odd years, since people first made their way to New Guinea, and low human populations as well as immense areas with no people at all meant they could always find pockets of forest to live in in peace. However, in the late 1800s the couturiers of Europe discovered the exquisite feathers, and began the trade in them for hats and other adornments that was to see hundreds of thousands of feathered skins transported across the world. Shipped to Europe with their feet removed for ease of packing, these were birds, so rumour had it, that had no legs because they never perched, remaining aloft in the firmament for their whole lives. Although heavily regulated now, the trade has continued in various guises right up until the present day.

Birds of paradise are obviously the number-one prize for any bird collector, but the native New Guineans have always used them to adorn their headdresses and supplement traditional bride prices. How do we tell them that, because of our past actions, they can't take them any more?

Handicapped by their lush feathers to be slow and blatantly obvious in the forest dark, these birds are victims of evolution. They display for their females in the same places annually, often clearing patches in the forest canopy so that sunbeams will spotlight them. They couldn't be easier to catch if they fell out of the sky on to village dinner tables ready-stuffed. Add to that the Indonesian government's shameful strip-mining of the Irianese forests, and their conservation will soon be out of our hands. The birds of paradise that remain are the last of what may already be a relic population, and are unlikely to survive.

While my sighting of the famous *cenderawasih* was very limited, a little later we got a very good look at another of Irian's airborne wonders, the *mambruk*, or Victoria crowned pigeon, the largest and by far the most beautiful pigeon in the world. The first one we saw seemed to be showing off for us, flying ahead down the river to perch in a tree until we arrived, before taking off again as we drew near. We saw the bird in two different colour morphs: jet black with blood-red trim, and vibrant indigo with dark blue and white. Both had bright-red eyeballs. The *mambruk* also has on top of its head a spray of feathers like the seed parachutes of an autumn dandelion.

Some of the most dramatic creatures of Papua, though, are marsupials – tree kangaroos and cuscus, feather-tail possums and bandicoots. Many marsupials live on plant sap, nectar or insects, but the quoll is a particularly vicious little creature that feeds on birds, reptiles and small mammals. About the size of a domestic cat and with big dark eyes, its exceedingly cute looks hide a ferocious demeanour and savage predatory instinct.

Irian is also the possible home of the most frequently seen extinct species in the world which, if it does still exist, is the largest carnivorous marsupial. The thylacine, or Tasmanian tiger, is/was an incredible creature, whose most novel feature is a kind of flip-top head, with the capacity to open its maw so wide that the upper and lower jaws form a single parallel trap of fierce teeth. During my visit in 1997, the newspapers were abounding with tales of a lone male having been spotted up near Jayapura on the boundary with New Guinea. Quite something, as the last bona fide recorded specimen died in Hobart Zoo in Tasmania in 1936. Tazzie remains totally obsessed with the beast, and people still wander its wilderness hoping for the unimaginable scientific kudos of finding living specimens. Sightings are recorded all the time, and in 2010 some ropy video of a possible thylacine was filmed at a distance. However, the quarter-million-dollar reward offered by a TV company for real evidence will probably never be claimed.

We saw kingfishers in the mangroves, and a variety of waders in the river mud, which sparkled with colourful fiddler crabs and spluttering mudskippers. The most conspicuous bird was the hornbill, with its gaudily coloured face and hollow horn topping its toucanesque beak. The arrival of this mammoth beauty is always announced by the majestic sweeping sound of its wings, like the swoop of slow-motion helicopter blades, audible when the bird is still well out of sight.

6

The jungle rivers have a personality more dramatic than the forest around them. While the forest seems largely indifferent to your presence, the rivers are psychotic, bipolar creatures that dissemble and deceive. Their café au lait depths hide dragons in the form of giant crocodilians and serpents, and demons in the form of water-borne parasites and diseases. But it's their furious roar that is truly terrifying, their capacity to slumber, lulling you into a feeling of security, then swelling with the tumultuous forest rains into a tyrant flow raging about your uncomprehending ears. Their flash floods have the power to strip you of all you own – your boat, your possessions, even your life.

We travelled several hours without any sign that humans had ever passed this way, and just as I was starting to feel that swelling pride and the accompanying fear that this was uncharted territory, some ramshackle huts appeared on the riverside. On spindly stilts and with tattered roofs of palm thatch, they were all deserted, looked as if they'd been thrown up in an hour and wouldn't last the week. These were the temporary shelters used by hunters and fishermen when they left their villages and went on foraging trips, a feature of their semi-nomadic lives. When you looked a little more closely, you saw that the overgrown land around them had been well cleared at many times in the past. Sturdy cultivated banana trees and sago palms stood around the shacks, a sign that far from being ephemeral overnight lean-tos, these huts had probably been here for generations.

The Asmat region is composed of several ill-explored subregions, named mostly after the tribes that inhabit them. After all this paddling, it seemed we should be getting close to my goal of visiting the tree-living peoples of the Kombai and Korowai, but, truth be told, we could have been further away than when we started for all I knew. There can be few places left in the world possessing so much of intense fascination that have seen so few visitors. According to my battered *National*

Geographic article, no more than twenty metres to the west of the mission village of Yaniruma, beyond the 'pacification line' where even the Indonesian army fear to pass, lie the 'Stone Korowai'. These tribes of warlike tree dwellers have had no real contact with the outside world, and they attack with poisoned arrows and spears anyone who ventures towards their territory. To the east, between the Dairum Hitam and Eilanden Rivers, are a score of tiny settlements where the natives live as much as twenty metres up in the jungle canopy, run naked, and greet visitors with suspicion and wonder.

Villages such as Yaniruma were purposely built by missionaries and government agencies to attempt to lure the Kombai and Korowai people out of the trees. Such places are completely alien to the natives, who are used to living in territories that have belonged to their clan for generations. Inside these territories the people live in households of patrilineal descent, in groups of tree-houses that may number as few as two or three for one village or clan. The people are traditionally polygamous and exogamous – that is, they take their wives, of which they have many, from external communities. Marriage is a massive undertaking, involving elaborate transactions and bride prices.

From Binerbis, Basman or Yaniruma on the central Dairum Kabur River it looked like it would be a day or so's walk to the nearest Korowai villages. Unfortunately, my map was woefully inadequate for the job of navigating up these rivers. We came to sizeable tributaries every few hours, any of which could easily have been the route we needed. Thing is, I really hadn't figured that the navigation would be down to me, assuming that my local companions would know the way like the back of their hands. This was not my first or last mistake, by any means. We were into the fourth day of our paddle when we finally, tentatively, branched on to a smaller river, which quickly narrowed. Increasingly we were having to get out of the canoe to tug it through shallow water, or to heave it over trees that had fallen across the river. We barely saw the sun at all that day, as the forest canopy hung right over our heads and hid the sky. For the first time, my two companions were talking to each other. Well, more arguing than chatting, really.

Drifting along that night in the pitch black, it suddenly seemed as if the riverbanks were lined with Christmas trees, draped with dense curtains of blinking white fairy-lights. They cast dramatic flickering shadows over the tangled forest canopy, shimmering off the water like that surreal bridge scene in *Apocalypse Now*. The glow was so powerful that you could have read a map or a novel by it. The first time I saw it, I jumped up in excitement, imagining that we had suddenly found ourselves an illuminated town in that Heart of Darkness. However, as the boat drifted up close, the billions of fireflies all became clearly visible, making a Mardi Gras of the hitherto hostile-seeming forest.

Fireflies are not actually flies at all, but beetles. In some species the caterpillar-like larvae and the females also glow – these are the glow-worms. In the last few segments of the abdomen, a chemical called luciferin acts with an enzyme called luciferase to produce a greeny-white light, one of the most perfect, unbelievable sights in nature.

I have seen fireflies and glow-worms all over the world, including in the UK, and they are always a true wonder. For American kids, a jar of fluttering 'lightning bugs' at their bedside was once an essential magical element of childhood. But I have never seen any, before or since, that have come anywhere near the show put on that night. Their sheer numbers and density made every other spectacle of bio-luminescence pale into insignificance. Added to that, each individual insect glowed with greater intensity than I'd seen before. But the thing that really sets this spectacle apart is that it is synchronised. Several hundred thousand fireflies will blink at exactly the same second, then perhaps another few hundred thousand will blink in unison before the first lot blink again. Eventually, in a process known as phase synchronisation, every single insect in a tree or along a riverbank will be flashing at the same time.

Night on the river, though, was rarely this tranquil. One black evening, as the dugout drifted close to the bank, our meditation was broken by a terrifying sound. An ominous, guttural roar, an unmistakable rumble, then a huge SWOOSH in the water and a violent splash, just metres off the side of the boat. We had disturbed a very, very big male saltwater crocodile. From the sound and the volume of the splash, I would guestimate he was at least four metres long, and probably more like five. A croc like that would weigh much more than all of us and our boat put together – potentially a metric tonne. He could have swatted us like a fly into the black waters.

Next day, late morning, and we found ourselves paddling into one of the great wildlife spectacles of Australasia. Suddenly, above the trees a vast flock of dark birds took to the skies, huge wings beating in slow motion as they whirled off in every direction. And then what had been just silhouettes poured from the branches like crude oil – hundreds, thousands, tens of thousands of small animals. These were not birds, but bats, great or Bismarck flying foxes, which are fruit-bats. Hanging during the day in a colossal colony along the riverbanks, once they took to the air they had a scary, primeval majesty about them, cluttering the heavens with black vampire shapes. Those that remained in the trees chattered, screeched and squabbled noisily as, cloaked in their wing membranes, they hung upside down from the branches by just a toe or two.

Though the whole scene had the distinct feel of a horror movie, these bats are worlds away from the actual Latin American vampire bats. They don't echolocate, feed only on fruit, and look more like possums with wings than most insectivorous bats. In fact, there was a school of biology (now discredited) that believed they were more closely related to lemurs than to micro-bats, and when you see a furry fruit-bat hung upside down from a branch, with a fig in its mouth and a long lapping tongue, you realise why.

That night we made camp on a flattened patch close to the water, as the lack of rain reduced the likelihood of flash floods. As we pitched the shelter and lit the fire, Michael and his son sat huddled together, their worried conversation conspiratorially secret from me. After we had eaten our sludgy stew, with me fervently dreaming of pizza and Guinness, Michael asked if he could borrow my map. This was the first worrying sign. They then sat well into nightfall poring over it in the firelight, turning it upside down and sideways, pointing things out to each other and arguing in an extremely anxious fashion.

I was by now getting more concerned. After all, my map took up just two pages in a photoguide, and showed the entire Asmat region. The river we were currently on was probably not on it at all. In fact, this was kind of the equivalent of trying to find out how to get from Chelsea to Dagenham on a half-page map of Great Britain. Excepting, of course, that if you get lost in Chelsea the locals are unlikely to eat you (although they might make wounding comments about your shoes).

Later, gentle rain pit-pattered on our tarp, singing me a sweet lullaby that sent me off swiftly to la-la land. Through slightly manic Lariam-fuelled dreams of evil cannibals dancing around the fire on which was set my own cooking pot, I suddenly became very aware of the soup I was being boiled in – it wasn't a *warm* wetness, but unpleasantly, chillingly cold soakingness. I woke with a start to the roaring of the river. The corner of the tarp had come loose in the rain, and was directing a constant stream straight into my sleeping bag. As I leapt up to retie the flap, I was suddenly overcome by how much louder, how much closer, the river seemed. Scrabbling around in my sodden pack, I found my torch, flicked it on, and swung the beam towards the flow. The water was lapping at the base of my tarpaulin. Brown, flecked white with foam, it had burst its banks and had an authority and purpose about it that it had never shown before.

I shook Michael awake, and we frantically began throwing all our kit into the bushes beyond the camp and taking down the tarp, all by the light of my crappy torch, clenched in my mouth to give me both hands free to work. Suddenly my stomach flipped over – 'The boat!' It had been tied to a piece of deadwood that was now totally submerged. 'Please God, don't let it be gone!' What would we do if it had been dragged away by the flow? It was still there, though, but dragged half under water by the flow. We plunged straight in – me oblivious to the fact I was still clad in my only dry clothes – and grabbed hold of one side, clenching our teeth and roaring with the effort. Our tree-trunk dugout, filled with water, would have weighed half a tonne or more, and the only way to get it out was to allow the water to flow out and empty the rest with our scooped hands. I noticed with dismay that my paddle had been swept away. It seemed at that moment that our lives were dangling by the threads of fate – every tiny mistake could turn our jaunt into a survival situation.

The only flat ground had obviously been made that way by flood waters, and wasn't deemed safe, so we slung the tarp up over a sloping bank covered in spiky bushes and sodden ferns, pulled all our stuff in under the shelter, and hugged our knees to our chests. I crawled into my soaking sleeping bag and could feel the water running beneath me dragging the remaining warmth away from my body. Even now, having done big Himalayan expeditions where I've slept out in minus thirty at 7,500 metres without a sleeping bag, I can honestly never remember being that cold and that miserable.

Near-hypothermic in the tropics – bloody perfect. While the other two dropped off to sleep, seemingly oblivious, I sat and shivered, and didn't get a wink.

The next morning, and Michael and son were a lot slower than usual in rousing themselves and getting going. Up until now, the sun had broken on a cheerful crackling fire, with a pot on for me to make tea. This morning found them fussing around the canoe and casting furtive looks back at me. The thought that they might disappear without me shaking the sleep from my eyes had me up like a shot. They weren't about to ditch me just yet, though, having decided to try whingeing tactics first. Unfortunately, it was all too bloody obvious that they didn't know where the hell we were, and they were getting frightened, particularly after last night's little experience.

I was furious. This had been my big adventure. I had suffered one huge false start already and months of dreaming and planning, and now it was petering out miserably. But my fury was completely pointless. They just went grumpy and left me to make breakfast on my own while my sleeping bag dried out in the sunlight. When it came to setting off, there was a huge row about which direction to take.

'Mr Istiv, it's better we go this way,' pointing back the way we'd come. 'We go back to Senggo.'

'No way! We've come all this way, I'm not turning back now.'

'I have to get back to Senggo, I have to bring in the sago for my family.' Strange that this hadn't come up before.

'No Michael, I have to go to the Korowai for my job. If I don't I will have big trouble with my bosses when I get back to my country.'

'This way very dangerous, you don't understand.' He looked genuinely worried here, and I didn't doubt he was really scared, though of what I couldn't be sure.

'I've paid you to take me to the Korowai, we've done four days of paddling, we must be nearly there now?'

'No-o-o, it's very far, many many days. You don't pay us enough money.' He was trying to be fierce, but his face wasn't quite convincing.

I pushed home my advantage.

'If we don't go to the Korowai, then you haven't kept your word. I won't pay you any more money at all!'

If only I hadn't given them half as an advance, this could have been an even better bargaining tool.

'You give us more money. One *million* rupiah.'

Aaaaarghhhh! I *knew* this would happen. Now I was really over a barrel, out here in the middle of nowhere and totally dependent on them. I couldn't exactly just sling on my backpack and storm off, could I?

'Michael, do you even know where we are? How far is it to Yaniruma from here?'

He waved off vaguely upriver. 'Five days.'

'Five days? It was five days five days ago! *What?* Are we going backwards?'

There was no answer to this. Michael started talking conspiratorially in his local tongue to Aguus, who nodded seriously. I couldn't have them plotting God only knew what behind my back.

'I tell you what, you get me to Yaniruma, Wangemmalo, Binerbis, any decent-sized village. Then you can leave me there and head back to Senggo.'

This was a bold and perhaps daft plan, but perhaps I'd be able to convince them to stay once we were there, or maybe I'd be able to raise more help from someone else. The two gave each other a long, knowing look, making up their minds.

'One million *rupiah.*'

Inwardly I sighed to myself. What could I afford? I'd agreed 300,000, and given them half up-front. I had less than a million left in total and needed to get back to the coast and away to a big town with a bank. That could easily take every penny I had. Any extra I gave them made it more likely I'd be simply unable to get out. An awkward, in fact an impossible, situation.

'I don't have a million *rupiah*, Michael, I don't have it. That is not going to happen. Listen, I will give you another 100,000. That is too much money, it is more than I can afford, but I will do that for you because you are a good man, and you have worked very hard. *Please*

Michael.' I tried not to let my desperation come across, but desperate I was. Please Michael.

No answer.

'OK, listen, let's continue up here for one more day, then if we don't find anything, we'll turn back. How does that work?' This was a great deal, an extra 100,000 *rupiah – and* not having to fulfil the promise. Throw a guy a bone here!

'Very danger this way. Better you have big Johnson.'

Oh God, don't you start.

We slogged on in silence for a good couple of hours. It's claustrophobic paddling on these smaller rivers. The sun is obscured by the forest canopy, and every hundred metres you have to jump into the brown sludge to tow the canoe over an obstacle, then five minutes later you find a fat leech hanging off your big toe. Additionally, we couldn't help brushing against the riverside vegetation, dislodging vast clouds of midges and other bugs, in numbers untold. They got in our ears, hair and even into our lungs, and half my time seemed to be spent coughing bits of flies out of the back of my throat. The air was thickly humid. Great blackheads and spots were burgeoning under the skin of my shoulders, and I could feel my face and nose turning as greasy as a teenage McDonald's chip server's. I'd taken Michael's paddle, and he had fashioned for himself a long punting pole, which he used to push us along. It altered our rhythm, but still worked reasonably well.

It must have been about eleven o'clock when we rounded a corner in the river, and there it was. In front of us on the north bank was our first tree-house. It was more substantial than the hunters' temporary homes we'd been passing, with the ground around it totally cleared and several dugouts tied up at the water's edge, some sunk and clearly useless, the others with a paddle lying across the sides and perfectly serviceable. The thatched shack stood above the riverside mud, as high as a two-storey house back home and set back perhaps fifty metres from the shoreline, on its own. It was in no way what I'd been hoping for, or expecting. The pictures I had seen in the *National Geographic* were of houses twenty metres (maybe five storeys) up at the very top of spindly trees, a true miracle of primitive human ingenuity. This was without doubt a tree-house, and would have been really impressive had my expectations not been so high.

It looked completely deserted. As we pulled up alongside the house, however, I could see someone inside. It was a woman with a shaven head, holding a mangy-looking little dog to her saggy dugs. There were brown strings wrapped around her throat and two rows of knobbly round scars – probably fashioned when she was a young girl, by burns or a bamboo knife – and a faded blue tattooed necklace below her clavicles. She looked out at us from the darkness in silent amazement.

The woman eventually tore her eyes away from me, then turned to her right and stammered a few words to someone in the dark, out of our view. The language she spoke sounded like someone impersonating the cluck of a chicken. Then, out of the darkness stepped a man. He was naked other than a string around his waist, and six-inch whiskers protruded upwards from piercings in his nostrils. Traditionally, these are either cassowary feathers or a bat's wingbones, but it was impossible to tell which. In my romantic imaginings he was supposed to charge into his hut and come out screaming bearing a bow and arrow or a fourteen-foot spear. Instead, he just stood and gazed blankly at us, looking as stunned as the woman.

Michael shouted a few words to him, but he just looked at him with empty, strangely milky eyes, clearly not understanding. There were a few moments of silence, before the impasse was broken. I reached down into the bag that the gifts were in: tobacco, some fishing lines and hooks, pens for the kiddies and the odd T-shirt. With a sizeable packet of tobacco in my hand and my heart in my mouth, I stepped out of the boat into the mud, waded up to the ladder and climbed slowly up on to the house's porch. I made my way up very easily, never taking my eyes off the man up on the porch, aware that the *National Geographic* guys were never allowed into a tree-house because the inside was considered sacred to the clan.

Standing in front of the man, with a clear view of him and his spouse, all my nerves vanished immediately. He stood under five foot tall, and was thin in the arms but with a large belly. He gave off a sickly-sweet odour, and it was soon evident why. His skin was practically coming off in clumps, mottled pink speckling the natural dark brown of his flesh, and the tops of his thighs and his genitals seemed as if they were filled to bursting with fluid. This was the elephantiasis that Ken had been telling me about in Senggo, along with a serious skin disease I couldn't identify as well as a bit of malnutrition signified by the distended belly contrasting with the skeletal arms. Really, food would have been a far more appropriate gift than tobacco, but that was the gift I had become accustomed to giving, and besides, I was up here now. The tobacco I pressed into his hand with a deep bow, and then stepped back.

For the first time he took his eyes off me, and looked down at what he had in his hand. He looked at the parcel uncomprehendingly, and then back at me as if for some explanation. '*Untuk merokok pak*,' I offered, 'for smoking.' No reaction. I tried asking some questions about the house and about his wife, tried greeting her and waving. Nothing. There would have been no problem just walking inside and checking out the inside of a Korowai house, but I could think of a million reasons not to, few of which make me proud. My feeble medical kit couldn't have begun to address the problems the two of them were evidently suffering from. There was nothing left to do but get back in the boat.

'We go back Senggo now, *tuan*?' Michael asked.

'I guess so Michael. Yeah, let's go.'

The trip back downriver was considerably quicker than the journey out, and while I was driven onwards by the urge to get back to what by now seemed like civilisation, that desire was undercut by the disappointment of a journey that despite all the effort had been a failure. On the main river I managed to flag down a motorboat, and paid them a seemingly obscene 50,000 *rupiah* to give us a tow on the end of a rope for a few hours. That took at least a day off our paddling time, but money was now a real worry, and I'd have to get to the coast quick, or risk an unimaginable cashless stranding. Once back in Senggo, the only way out of the Asmat region was from Agats on the south coast, from where I hoped to jump on a cargo ship heading somewhere – anywhere – else. It would have taken a month to paddle all that way, but I figured I would probably be able to get on to an engined boat going down the main river. We'd only passed two or three in our week's paddling, but it seemed likely they'd be more regular between the two larger settlements of Senggo and Agats.

Some explorer *I* was. My big adventure was done. It had been a spectacular failure. In addition, it turned out to be an entirely unreasonable extension to my working trip, as *Rough Guides* eventually decided that as no tourists ever went there, they'd sub my contribution on Irian Jaya down to twenty-eight pages. So the world's second-largest island got as much space in the guidebook as a single town in Bali. I felt like Ford Prefect when his twenty years on earth were summed up in *The Hitchhiker's Guide to the Galaxy* with 'mostly harmless'.

7

It was 2009, and thirteen years since I'd set foot on New Guinean soil – a return seeking some kind of redemption. This time I had the chance to explore places no Westerner had ever seen before, to discover species new to science, to instigate the protection of environments that might otherwise perish beneath the logger's chainsaw ... But I'm getting ahead of myself. A frightening amount of water had passed under the bridge since I left New Guinea with my tail between my legs. Please bear with me while I attempt to pack a huge portion of my life into a few paragraphs.

It was the late nineties. Writing for the *Rough Guide* remained enticing for only another year, before the allure of the job title was no longer enough. It wore off as I came to realise that the only way I could continue to be a writer while also paying the bills was to have second and third jobs in bars and nightclubs. The stench of stale beer and cigarettes, though, was nothing compared to the bitter taste of my wounded pride. I found myself wanting to say to every customer, 'That'll be two pounds fifty please – and by the way, I am a published professional writer, this bar job is just temporary.' I needed a new goal, and once I'd decided, I'd have to throw absolutely everything at it.

That goal was television, and the field was packed. The way I saw it, my only chance would be to come at it from a different angle from everybody else, and give it every ounce of my tenacity. I was no better than anyone else in the field, and had no experience whatsoever – but I'd want it more, be prepared to sacrifice more, make myself stand out by being a bit different.

Then, in a truly bizarre streak of luck, I met a girl in a London salsa bar who offered me a job modelling home fitness equipment on QVC, the shopping channel. At the time I had precious little pride left, and it offered me about two months' wages for each weekend spent grinning and gurning on a treadmill, while some loudmouth steroid-fuelled American yelled at people to buy his stuff. It turned out that just a few of these ludicrous jobs got me enough money to buy a cheap home video camera and a plane ticket to the most exotic and frightening place I could think of. Back in the late nineties, this was Colombia, a country on the brink of civil war, plagued by violence, drugs and lawlessness, but also possessing snow-capped Andean peaks and untrammelled rainforests.

Colombia really lived up to its reputation. I went out there with a friend who'd only ever really been to Spain before, and he promptly nearly got killed by drug-runners and corrupt police before getting arrested and almost deported. Coming in to a police compound at four in the morning to find him handcuffed, pistol-whipped and surrounded by a ring of police with machine guns levelled at his head threatening to shoot him and dump him in the jungle, remains one of the most surreal and terrifying moments of my life. However, I did manage to spend a good deal of time in the jungle, catching highly venomous snakes and scorpions on camera, filming the whole thing in a warts-'n'-all kind of style. I had to do it on my own though, as my friend got heli-evacuated from the jungle, near death after suffering every kind of affliction imaginable, due to deliberately ignoring every piece of advice I gave him. No wonder I am so much happier travelling solo. Once home, I linked two video machines together to edit the film and copy it hundreds of times, then sent the tapes to just about everyone working in television in the UK, along with a proposal for a series entitled '100 Things to Do before You Die'.

This may sound impossibly hackneyed now, but in 1997 I believed I'd hit on something unique. The fact that literally hundreds of books and TV programmes have been made to this format since suggests I really had hit on something, but didn't have the clout to make it happen. After months of let-downs and nearlys, I was called in to the National Geographic Channel in west London to meet with their vice-president, an Australian ex-presenter called Bryan Smith. Bryan is a spiky and intimidating character, tall, red-headed and physically imposing, and as I was eventually to realise, something of a TV genius. He had a typically Australian ability to call a spade a bloody spade, and didn't trouble himself with trivialities or polite niceties, but made big calls with a forthright decisiveness that it later dawned on me is what sets truly important people apart from the likes of me. All of the most valuable things I've learnt about television I picked up from Bryan in that first half-hour meeting, things I keep with me even today.

Despite the fact that I was clearly not even remotely qualified for it, Bryan hired me on the spot as not just a presenter but also as producer. He even gave me an absurdly grand title – my business cards read 'Steve Backshall, National Geographic Adventurer in Residence'.

Bryan put more faith in me than I had any right to expect. When I turned up for my first day's work, he merely gave me his company credit card and the address of a London production house, with instructions to buy a professional video camera and sound kit. There was what seemed at the time a sizeable budget, and the frighteningly open invitation to 'just go and film yourself doing what you do'. He gave me carte blanche to go on nutty expeditions filming myself, before editing the film wherever was financially viable. This meant extended stays in Tel Aviv, Sydney, Washington, even six weeks in an antiquated edit suite in a New Delhi slum with an editor who spoke no English and a computer that crashed every hour. Years spent wielding my trusty Canon meant I was a competent stills photographer, which gave me a head start on learning the filming side of things. But the editing, producing, research and presenting I really just had to pick up as I went along, or risk Bryan's wrath.

There was no time for complacency. In the year 2000 alone I travelled to thirty countries, catching giant anacondas and tiny venomous vipers, filming everything from wolves to great white sharks, crossing deserts, glaciers and rainforests. As you'd expect of a young man who'd never handled a professional video camera before, let alone edited or produced, the results were often just plain shocking. I filmed for entire days with the boom mic in shot, whole sequences with the sound switched off, and dropped a camera in the middle of the desert in the only (sink-sized) pool of water for a hundred miles – totally killing it. It was the ultimate sink-or-swim baptism of fire, and dragged me forcibly into the world of television.

It took probably a year or so, spluttering around in the deep end to relentless paint-stripping criticism from Bryan, before I managed to put together some quite credible adventure films, got nominated for several awards, and had simple shows I'd made entirely on my own doing pretty well in the ratings. After five extraordinary years at

National Geographic, I was flying high, loving life, brimming with the confidence of what I'd achieved and what I had ahead of me. But I was about to be given a whopping kick in the teeth.

I met with Bryan one day in a posh London restaurant to negotiate, I thought, a massive two-year contract. I'd spent days working myself up to asking for more money and bigger and better projects. Instead, I was about to be told that Bryan my mentor was moving on, and that the new boss didn't like making films with English-speaking presenters that would be broadcast in the 120 countries round the world National Geographic transmitted to. Having seen myself bizarrely dubbed in Hindi, Hebrew, Dutch and Tagalog, I had to admit his argument made a lot of sense. However, I was out on my ear.

It was a crushing blow, and I walked, dazed, out of the restaurant feeling like the bottom had dropped out of my world. Within two weeks, close colleagues from National Geographic didn't even return my calls or emails – even people I'd considered to be really close friends, had stayed at their houses and knew their partners and children, dropped me like a stone. That really stung, and taught me my most important lesson about the ruthlessness and fickleness of television. As a presenter, you are either the most treasured and worshipped commodity, or you are nothing – an embarrassing wannabe, a failure or a has-been. I'd never trust anyone or anything in the television industry again. The gravy train was derailed, I was going to have to get a proper job.

It's at times like these that you rely on a bit of luck and some very good friends. Unbeknownst to me, one of my old university pals was working for the BBC in London, and without prompting contacted the producer of the BBC Natural History Unit's longest-running kids' wildlife programme, *The Really Wild Show*, to tell her there was a National Geographic presenter 'you really should take a look at'. Not surprisingly, I was gobsmacked when the BBC actually called *me*, asking if I'd like to come down for a chat. Again, there was a sense of serendipity about my whole first trip down to Bristol to meet the *Wild Show* series producer Lucy Bowden. I took down a video of me on one of my self-filmed expeditions, and managed to convince her that I could save the series a fortune by going on the road alone, without cameraman, soundman or director, but still bring back the goods. Lucy was extremely bright, accomplished and impressed me enormously, but for some reason decided to take a chance on me.

Again, I walked out of my very first meeting in the office at BBC Bristol hired, with someone willing to put all their faith in my abilities. This position was an incredible privilege. First, I'd be working for the BBC Natural History Unit, the most prestigious wildlife filming institution in the world. Second, I'd grown up watching *The Really Wild Show*, with legends like Terry Nutkins, Michaela Strachan and Chris Packham, and Michaela was still working on it sixteen years later! I can remember vividly my first meeting with her, chatting blithely about her love life as if we were old pals, as being one of the most surreal moments of my life! Unlike the other presenters, who had loads of other work, I directed my own shoots and spent my time in between filming trips working in the office. To begin with, as I didn't have a place in Bristol, I took my little mountaineering tent and camped out in a caravan park down near the Avon, cycling in to work up through Hotwells.

Over three delightful years on the series, I never really gelled with the other presenters, though. Nick Baker, my counterpart, is a biologist, a world-acknowledged expert on British wildlife and the finest natural history presenter I've ever worked with. I, on the other hand, was painfully aware of my lack of scientific credentials, and felt self-conscious, almost a bit of a charlatan. Even some of the runners and researchers at the NHU had PhDs. I was just a bloke who knew stuff about animals.

I decided to set about remedying the situation as a matter of urgency. The first problem was that I'd never even done any science A Levels, so had to start from scratch. First I took a science foundation course, then a diploma in natural sciences, and then started on my biology degree with the Open University. Anatomy and physiology, animal behaviour, evolution, even endocrinology and microbiology, I buried myself in the lot, while still doing my full-time job as well as writing natural history books and articles. For many years I could count not only the days, but the hours and minutes I wasn't working. Every second of every day was spent either on the job or buried in books. Never having had a particularly logical brain, and in addition being cataclysmically flummoxed by mathematics, this was the hardest period of my life. I'd breezed through school and university the first time round without breaking sweat, but now found myself struggling – and even failing – in the academic world. One tough year-long module I battled through successfully, only to eventually fail as I had to be out of the country filming when the critical exams were held.

But though it was the toughest, it was also my most enlightening period, waking me up to a whole new world of knowledge and reason. It was as if someone had turned the lights on in my head, and suddenly all I wanted to do was immerse myself in science. I'd always been good at identifying animals and remembering loads of facts about them, but now I began to really understand them, and to see the bigger picture in everything I observed. More than anything, I read obsessively about evolution, and began to see in its teachings a religion more potent, exciting and joyous than any spiritual texts could offer. The miracle of how we and all the other organisms we share the planet with came to be here, filled me with wonder. I felt like Darwin himself, suddenly seeing the world with a child's eyes!

In the words of Christian biologist Theodosius Dobzhansky (1900–1975), 'Nothing in biology makes any sense except in the light of evolution.' I could equally say that nothing in my life really made any sense until I truly understood evolution. Before Darwin and Wallace came along, the entire story of creation was told in about a hundred words in Genesis, the world was about 6,000 years old, had been brought into being in its current form, and any more enquiry than that was considered heresy. The truth was that every single animal form or behaviour had evolved over countless millennia. There was no faith or belief required, as once you understood the science, it could all be *proved*. Perhaps my turning point text was Richard Dawkins' *Climbing Mount Improbable*. Not his best work, but in it he describes the evolution of complex organs such as the eye and the wing, step by step through examples that still exist today. I couldn't believe how much we actually know! There are almost no missing links left, the breadth and depth of modern human understanding is just overwhelming. Science is just so elegant and simple; it's all about certainties. Unless you can prove a theory is true through experiments, then it has no merit. And far from being restrictive, there were libraries-worth of words to be read on the subject, a thousand lifetimes worth of study and understanding – I had to get started right now! I felt like I had joined a privileged society, one that had an infinite amount of enlightenment at their fingertips.

There is no deep meaning of life, we are merely a vessel for passing on our own genes. Some may find that frightening, but for me it was liberating; we are not inherently special. We can make our own morality and life path, and there is no need to fear death. In fact, the only thing we need to fear is not getting every ounce out of this too short, wonderful, miraculous life. From there on in, every walk in the countryside was a learning experience. I guess this really is when my lifelong love of nature became an obsession. I buried myself in zoology, found a whole new group of inspiring naturalist friends, took up breeding exotic insects and reptiles, building up a regular menagerie at home. On expeditions where I knew I'd be forced to read whatever I had, I'd deliberately leave all reading material at home except for an animal encyclopaedia, which I literally learned cover to cover.

Three years in, and we concluded a series of *The Really Wild Show*, by common consent one of the best they'd ever done, nominated for a BAFTA and both critically and popularly successful. Again, though, the fickle, sometimes baffling nature of television was to rear its ugly head. The brand-new head of Children's BBC 'wasn't interested in animals', had her eyes on making a big statement and cut the series dead, after twenty years. Six months later she moved on to another channel, but children's wildlife programming at the BBC was for the moment dead. Once again my train had come off the rails, so I was faced with living hand to mouth, seeing the career I'd worked so hard for going up in smoke.

Looking back, this was a vital time for me. Two production companies from my National Geographic days unexpectedly offered me series producer roles, which would have been a big deal, putting me on the path to a management role in television in my early thirties. I was very torn. After all, the great David Attenborough himself actually spent time as the controller of BBC2. All common sense and reason told me to take one of the roles. However, neither of them appealed to me. It's always been the expeditions, the animals and the storytelling I've loved. I have no interest in television in itself as a medium, and certainly didn't want to be sat in an office back in the UK sending other people in the field to have all the fun.

Yet again, serendipity took a hold of my future. Steve Greenwood, a producer at the Natural History Unit, had been given the task of putting together an expedition series to Borneo. He'd been told that I had done several big expeditions there and knew the place well, so one of his first meetings in the very early planning stages was with me in the BBC Bristol canteen. The idea was to take a team into an undiscovered area of Borneo, to search for new species, film the existing animals, and look for a way to protect the forest itself. I brought along maps, books, photos – I had really done my homework. Steve offered me the role of assistant producer on *Expedition Borneo*, which is just one up from researcher, and below the level I'd been at even back when I started at Nat Geo seven years previously. All the advice I got from friends, my agent and family urged me not to take it, not to take such a massive backwards step. However, the idea of several months back in the rainforest on a project with such ambition was just riveting. I also secretly harboured the hope that I might be able to make myself indispensable and clamber up the ladder a bit by the time the expedition got started.

Sat in the canteen, Steve seemed an odd choice to lead an expedition into the Heart of Darkness. Tall and lanky, with fashionably floppy grey hair and glasses, he looked more likely to be giving classes on metaphysical poetry than taking control of a team that would include SAS veterans and the most experienced natural history camera people in the business. Steve had trained as an entomologist, then worked as a radio presenter on Radio Inverness; I took one look at him, and thought the jungle would give him a good kicking.

So we left the UK together to search for a basecamp site in some of the toughest jungle Borneo had to offer – and I was soon proved hopelessly wrong. Steve *loved* it. He was as happy as a Floridian at an all-you-can-eat buffet, endlessly entranced by the tiniest orchids and the most innocuous-looking invertebrates. He relished the grim marches, and in his rain-spattered specs would grin ear to ear – even though he attracted leeches like candy-floss attracts kiddies. We did an experiment with one particular leech; I slowly moved my finger closer and closer to it, and when it was about three inches away the leech sensed me and moved my way, looking for a blood meal. When Steve did the same, it was still over a metre away when it turned and started frantically shuffling after him. (I'm surprised they don't scent him when he's at home in Bristol.) He also turned out to have a razor-sharp mind when it came to a story, and I began to like and respect him enormously. The two of us took to organising our grand adventure like a couple of lads building rafts for a summer playing Swallows and Amazons. We explored much of Borneo before I chanced upon a little-known corner of Malaysin Borneo called Imbak Canyon. The first time Steve and I journeyed there, we knew we'd found our location. Remote, physically stunning, little explored, totally unprotected and vulnerable. This was our chance to achieve something more than a mere television programme.

By the time the main expedition came around, I'd managed to spirit into the itinerary certain onscreen jobs that only I could do: namely, the epic first ascent of the peak to the south of our basecamp, plus a descent into the bowels of the earth in the Mulu Mountains cave system. When the scientists turned up – mega-minds from places like Oxford University and Berkeley – I was frightened half to death that I'd be found out, and found wanting, by my more illustrious colleagues. However, within a day or two I found I could more than hold my own. In fact, sometimes my field experience outstripped their book and lab learning, and I even found myself telling the professors things they didn't know – things they repeated on screen!

Quickly my confidence flourished, and in the end I shared equal screen time with the other presenters in a series that was a great success, went round the world, got nominated for all sorts of awards; and most importantly in the aftermath of our work, our patch of forest was sequestered as a Grade One protected reserve.

The grimy, warts-'n'-all style of *Expedition Borneo* was to my mind nothing new. In fact, it was an echo of David Attenborough's very first wildlife programmes, the *Zoo Quest* series started some fifty years earlier. However, it had an honesty about it that television seemed to be lacking at the time, and was eagerly embraced, leading to a bit of a glut in *Expedition*-style television programming, not all of it very good. However the Natural History Unit took a more considered approach, committing to doing just one big expedition series a year, ensuring that they would be epic. Set in Irian Jaya, now known as West Papua, the Indonesian-occupied half of New Guinea, the *Lost Land of the Volcano* expedition was all to take place in the eastern half, Papua New Guinea. This second expedition would itself fall into two halves. The first I was not looking forward to very much, as I was to spend three and a half weeks down a cave system. The second offered me the chance to emulate my hero Alfred Russel Wallace, to search for new species of animals in as yet uncharted regions of the mighty island. The information we garnered we intended to place – forcibly, if necessary – into the hands of governments, NGOs and the general public, in an attempt to prove this place was unique and worth saving.

It was perhaps the purest objective I'd ever had on an expedition, and the most exciting, the chance to be part of a team that could make discoveries of real significance. No one knew for certain what we might find, but previous biological assessments in far-off parts of New Guinea had discovered not only new kinds of spider and butterfly, but bats, tree kangaroos and other marsupials, and it seemed likely we would do the same. We had assembled a team of frightening intelligence and experience, and were about to have the opportunity to achieve something substantial. It was an extraordinary privilege, and an expedition that many of the world's top biologists would have given their eye teeth to be part of, and I was one of the chosen! I still sometimes wake up in the morning expecting the whole thing to have been a dream.

The name Papua is derived from a Malay word for 'frizzy', referring to the hair of the natives. 'New Guinea' is a reference to the fact that these same people were said to resemble the peoples of African Guinea. After the First World War, this former German colony was taken over by Australia. During the Second World War, Papua New Guinea was the site of some of the bloodiest campaigns of the eastern conflict, with a quarter of a million Japanese, Australian and native soldiers killed in battles that ravaged the island's jungles. The country was granted independence in 1975, but even today receives most of its foreign aid from Australia and is quite closely tied to its previous colonial rulers. The head of state is still Queen Elizabeth II. This may seem incongruous, but it does allow the Duke of Edinburgh perfect opportunities to make his sublime gaffes, such as when, on meeting a

British student in the capital city Port Moresby, he commented: 'You've managed not to get eaten, then?'

Forty-five thousand years ago, New Guinea was contiguous with Australia, and would have been alive with megafauna that are sadly now long lost. Then, ten to twelve thousand years ago, sea levels rose, cutting off the New Guinea landmass from Australia and so separating the New Guinean peoples from the Australian aborigines. The woods were inhabited then by monitor lizards weighing more than a tonne that would have eaten today's Komodo dragons for breakfast, sixteen-metre-long pythons, wombats the size of hippos – called diprotodons – enormous kangaroos and crocodiles, and birds four times the size of the still gigantic cassowaries that are New Guinea's largest modern land creatures.

All of these animals were to be hunted out of existence when the world's most fearsome predator arrived overland from Australia. Man came to New Guinea just forty-five thousand years ago – a mere breath in geological time. The new arrivals surged inland using the biggest rivers as highways, and found their way to the highlands where they settled in the agrarian communities still found there today – making them the oldest human societies on the planet, and the most continuously sustained cultures ever known.

Men had already evolved into the physical form we see today, though were less developed in language and culture. They did, however, have sufficient communication and weaponry to wipe out large animals wherever they spread. North and South America lost their large land mammals such as the giant sloths, short-faced bears, sabre-toothed cats and mammoths perhaps thirteen thousand years ago, after the Clovis peoples had made their way with their spears and arrows over the glacial land bridges. In New Zealand the giant moa and Haast's eagle, thought to have been the largest eagle ever, disappeared just six hundred years ago with the arrival of the Maori. (In Madagascar the even bigger elephant bird – three metres tall with an egg a metre in circumference – and giant lemurs disappeared in the seventeenth century with the arrival of man.)

But it is here in New Guinea that the skill of man the hunter was perhaps most keenly felt. The immigration of sophisticated man led to numerous extinctions and even today, there is little in the way of sizeable megafauna. Nonetheless, although there are no large land mammals left, and the animals that did survive in the dark forest depths were left at low densities, they have evolved over millennia to fill every ecological niche known, creating an unparalleled diversity.

The island of New Guinea is a complex one, not just divided politically east to west but comprising several large islands as part of Papua New Guinea. Our first destination was to be the vast island of New Britain, situated in the Pacific Ocean off New Guinea's northwest shores. If New Guinea is a turkey about to drop an egg on Australia, New Britain is its swept-back wing.

The only Westerners who ever visit this extraordinary place are missionaries, many of whom have over the years ended up on some Papuan tribesman's dinner table and not as a guest. We, however, were here for New Britain's fabulous jungle mountains, and the unexplored cave systems that pepper them like holes in a Swiss cheese. Our team was the biggest we'd yet put together for one of the expedition's satellite trips – ten of us in total. I was to be the team naturalist and presenter, with our safety and ropes guru Tim Fogg to provide the main voice of reason as we pushed on into the darkness.

Tim has an interesting history. He has an industrial ropes access company – cleaning skyscraper windows and cooling towers, steeple-jacking, doing tree surgery and rigging ropes for television companies. He worked on the British Antarctic Survey for many years, and has genuinely been around the block. He is slightly taller than me and a little slighter in build, but could crush my hand in his like a walnut. He's wiry and sprightly as a Jack Russell, and with a contained energy about him that suggests he always has something in reserve. I can't quite believe that Tim is now well into his fifties – I don't know many twenty-year-olds who'd be able to match him for fitness.

But the big reason why Tim is the first name on every expedition team sheet is that he's the person you want on your side when things go bad. When Rudyard Kipling wrote the poem 'If' he was thinking about old-fashioned heroes like Tim, a man who truly can keep his head,when all around are losing theirs. Over the last half-decade I've had Tim at my side in some of the edgiest situations I've been through. As things get more and more frantic and everyone else starts to panic and make mistakes, Tim gets more considered, more measured, and thus has a calming effect on all. I idolise him, and would love to think that sometime in the future I might be a fraction of the man he is.

His wife Pam, also possessing huge amounts of caving and rope experience, was the sole female member of the team. Northern Irish in humour and temperament, she is tiny in comparison to the rest of us, and our secret weapon should we encounter any tight spaces during the adventure.

The cameraman was to be Keith Partridge, one of the most respected, and I think the best, adventure cameramen in the world. He's filmed such masterpieces as *Touching the Void* and *The Beckoning Silence*, but most won my admiration when filming me climbing alongside a thundering waterfall *inside* a glacier in Alaska. Climbing there, hammering my ice axes into bullet-hard blue ice, was one of the toughest things I've ever done, but Keith filmed the whole thing carrying a fifteen-kilo camera, hanging from a rope, twisted sideways and balancing in a splits position with just one crampon point holding him in place. Throughout the whole thing, as ice plummeted down about his head, he kept every shot perfectly stable. I later realised that in situations like this he holds his breath for several minutes at a time, so the shots don't go up and down with his breathing! With his squaddie haircut and lean runner's build, he looks like he could twist a cage fighter into knots, but it's his consummate calm ability in hairy situations that really impresses. With him and Tim on the same team, you feel as if you can accomplish anything – while also knowing that on the rare occasions they shake their heads and say No to something, you shouldn't try and argue!

The production team consisted of the two Jonnys, Keeling and Young, both very experienced BBC directors. There was also a selection from the original caving team who'd been the first people to visit the area, on a National Geographic expedition in 2006. 'Moose' is a lanky Brit of about six foot five, and a totally obsessive caver who barely seems to have ever seen the sun; Jean-Paul is slight but fearsomely strong, and huge fellow Frenchman Willi looks capable of carrying a Mini Cooper up a mountain. With their wonderfully strong Gallic accents and size disparity, they were the spitting image of Asterix and

Obelisk. Last was Dave Gill, his best caving years behind him, but ready to make sure that basecamp ran efficiently. He had been the leader of the original expedition, as well as the one who convinced us the trip was achievable.

8

The gorge's sides were near-vertical, and the gorge itself appeared so narrow that we could have leapt from the skids of our chopper and caught hold of the dangling vines on either side. Hundreds of metres below us ran the white-water river that had chiselled this mighty gouge out of the soft limestone; above us a densely forested shelf, then mountains. We zipped around the twists and bends in the gorge at breakneck speed, every few minutes in the air representing a year of exploring and discovering. But then our pace started to slow, and as we rounded one last corner we were faced with one of the Unknown Wonders of the World. The leviathan cliff face before us was punctuated by three holes, the mouths of caves that penetrated deep into the limestone. Out of these caves gushed frothy white waterfalls, dropping a hundred metres down to the mossy plunge pool far beneath. Our mission would see us descending into these caves like spiders on silk, then heading into the darkness, taking the first ever light into untrodden passageways.

Even getting to the first cave's entrance was to take the best part of a week. First step was to charter a plane and fly into the small New Britain village of Pallmallmall. Despite having a runway big enough to land quite a sizeable plane, Pallmallmall was only home to a few hundred people. We stayed the night right by the coast, in the threadbare and mosquito-ridden Cocky Lady's Guest House. Illuminated at night solely by candles and our head-torches, the guesthouse's walls were papered with AIDS awareness posters, some in the form of crude cartoons, others in stark appeals from Papuan rugby league players to use condoms and have only one partner. In a village without power, which had not seen visitors or a supply drop for months, we rocked up with several tonnes of food and expensive electrical equipment, and provided the villagers with a spectacle that would probably be spoken of in their folklore for generations.

Next morning we headed to the quay to await a helicopter that would take us the final half-hour into Ora, the closest village to the caves. As we sat on the quay surrounded by towering boxes of tinned tuna and quizzical-looking locals, gazing out at the emerald mountains across the bay in front of us, suddenly the sea erupted with dolphins. These were Pacific spinner dolphins in a pod of at least a hundred, feeding and playing in the bay. As the noisy whirlybird carried its cargo of cavers and diabolical food up towards Ora, the director Jonny Keeling and I swam out into the warm clear sea and floated blissfully as the dolphins leapt out of the water, spinning on their axis as many as five or six times, occasionally turning somersaults or just landing flat – a bit like the cheeky fat bloke who deliberately belly-flops at the local pool. A couple of times they did super-close fly-bys, and looking down with our masks on we could see them streaking past us, our ears ringing with their clicking and squeaking chatter. This seemed to us the most elegant of good omens. Finally the kit drops were all done, and it was our turn to climb into the helicopter for one of the most spectacular journeys imaginable.

The whirring machine bore us first over the dark-green bubblewrap of pristine jungle-clad hills, then over knife-edged ridges and massifs, before dropping into the gorge, the white-water river raging below us and puffballs of mist teasing the trees. Even two hundred metres above the water, the gorge was so steep-sided and narrow it seemed the rotor blades must surely be clipping the leaves on both sides of the valley. We passed many waterfalls alongside us, but none were worthy of mention after we had seen our destination. For suddenly we rounded a corner, and there was Mageni cave. A vertical green cliff stood several hundred metres above the river, and erupting from the rock, about two-thirds of the way up, three magnificent waterfalls. The central and highest of these – double the height of Niagara – marked the mouth of Mageni, a dark cave with one of the world's most spectacular waterfalls, dropping down from it to the aquamarine plunge pool below. This is where we would be spending the next sixteen days.

Landing at Ora village in our chopper made me feel uncomfortably like an ignorant pith-helmeted colonialist. The last time I was in New Guinea I shared some of their language and lived amongst the people, sleeping in their huts with them and their pigs and their pigs' fleas, yet despite having nothing more technologically advanced than a few biros, they reacted as if I was some kind of peculiar alien. How, then, must these villagers have looked upon us now, whirling down out of the sky in our aluminium flying machine, then bursting out with our huge high-definition cameras rolling and proceeding to address them awkwardly in our own tongue.

Minutes after we arrived we wandered down with a huge entourage to the hut of the village Big Man, where we squatted down for *tok tok* (talk). Johnny Young, the other director, wanted me to quiz them about how they felt about global NGOs working to turn their area into a world heritage site. The translator informed me they didn't understand what I meant by 'world'. Sat in one of their dark huts, with them eyeing us silently and looking confused and us blasting Super Trooper lights everywhere to film their weirdness, I felt utterly ill-at-ease, like I was some imperialist fool who'd bumbled thoughtlessly into their lives without bothering to learn anything about them. On my earlier trips I'd studied the local language and could hold a decent conversation, but this time round I didn't have even a few basic words of pidgin. In my mind this trip had been all about the cave – I hadn't even contemplated an encounter with the people who owned the land. The situation was slightly salved by the local missionaries, who had been in Ora for close on a decade and spoke the language. They were good people, an Australian family with two delightful, intelligent children, who had done great things to educate the village and provide medical aid.

Next morning, we gathered our luggage into porter-friendly loads, divvying out the bags amongst the queues of locals desperately hoping to get the chance of a good wage from a day's carrying. As so often in these situations, it took several hours to get through the procedure of making sure every load was accounted for, and that excessive loads weren't being taken by children or old men. In classic New Guinea style, the youngest and strongest men would push to the front of the queue, put their names down in our book of porters to be paid, and pick up a nice fat load. They'd then carry it off to their wives, dump it down in front of them and amble off for a smoke. We had to intervene when we saw our generator (by far the biggest and heaviest load) about to be hefted by two lads no more than ten years old!

Eventually, though, the march began, and in glorious sunshine. Making our way between the huts, we crossed the village, climbed a small hill and entered the forest. This was the last time we were to see sunlight for several days.

It was no more than a two-hour yomp into camp, and we were all thoroughly enjoying the physical exercise until it started raining – full-on, torrential, world's-about-to-drown rain that seemed to want to beat you into the forest floor. Luckily I had all my stuff in a waterproof rucksack, or every single item of clothing would have been soaked for the entire next three weeks. Some of the other guys were not so lucky. For the last fifteen minutes we were slithering through the mud, all the time looking at the watches and thinking: Must be there soon, it'll just be around this corner.

When we finally arrived, though, it was to find about thirty Papuan porters clustered under a hastily erected blue tarpaulin on the spot of cleared land we had chosen for our camp, obviously not wanting to leave while the rain was still hammering down. This was fair enough, but unfortunately it meant sixty bare feet – eighty counting all of us – tramping the floor of our camp into knee-deep mud – it was like Glastonbury, but without the music and the hippy girls. After a few hours, the rain still not abating and gallons of water pouring off the tarpaulins into the mud, we were more than a little uncomfortable.

In fact, this simple turn wreaked by the weather gods was in danger of bringing our whole expedition to a halt that it might not recover from. It certainly ensured that for the rest of the trip we would never be dry, not for a second. The ooze would penetrate everywhere. Without a dry centimetre to put our valuable filming kit down on, and with rivers running right through our sleeping area, the squalor was inescapable. Whereas our clothes might eventually dry out, you can't really hang a high-definition camera out on the washing line. The team picked up every prospective makeshift tool and began to scoop buckets of mud off the floor. I took a pick and shovel and started to dig a moat around us. But it filled faster than I could empty it. This thick, squelchy mud was with us for the rest of the trip, and turned every step into total misery. It was going to be all downhill from here.

In fact, I hadn't been looking forward to this part of the expedition at all. First, I'm not big on caves, which has some implications when you're about to spend three weeks down one. I never had much of a problem with them until television work forced me into some of the gnarliest to be found. After getting properly stuck a fair few times, I'd all of a sudden started to develop a bit of claustrophobia that had never been there before. In tight spots, I actually began to feel myself panic, and had to force myself to breathe easy and slow my pounding heart rate down. Second, the expedition was going to be a nightmare for us to film. With just one cameraman and no soundman or lighting person, we'd be battling against every challenge known to television: mud, heat, wet, and their bastard son humidity, that soggy sprite with a wicked sense of humour that gets into the workings of anything electrical. In the rainforest, cameras just seem to decide they've had enough of filming. Caves are even worse. So making a film in a cave system in the rainforest was going to be harder than trying to film snow leopards doing the Macarena.

That evening, we got together for our first war council. Despite being a team full of natural leaders, everyone seemed fine with a democratic approach to the problems ahead.

'Our first issue is where we drop the ropes down to the mouth of the cave,' said Tim. 'What's going to be the safest way of getting down there, guys?'

'Perhaps three hundred metre like this,' said Jean-Paul in his very French English, pointing off into the forest, 'the ridge is fall down WHI-I-ISHT!' – he makes a sound effect like a sword swooping the air, and his hand cuts downwards. 'We drop the ropes here, then reroute maybe three maybe four times, like this' – now he makes shapes with his hands to show slack-hanging traverses.

'Is that out on the open rock?' I ask.

'Not really, no, Steve,' says Moose, in his thick Derbyshire tones. 'It's through nasty green vegetation. Bit like climbing through a shrubbery.'

'That's not going to work great for us on camera,' I remark. 'The ideal would be if we could drop right down on to the cave mouth. That's the glory shot.'

'Yeah, you'd never do it like that, though, if this were proper caving,' Moose reasons. 'Doesn't make any sense to drop down on to the top of the waterfall like that – don't forget we have to do it every day for two weeks.'

'I know, safety has to come first, but we're really missing a trick if we don't get that shot – hanging out over the falls, splashing your feet down above the drop. Imagine the shot from the helicopter!' I'm getting carried away now.

'Maybe we can set that up later,' muses Tim. 'We have two thousand-odd metres of rope.'

'Zees you will need, ah?' says Jean-Paul with a shrug, in a manner too French for words. It sounds like a question, but is without doubt a statement. 'There is many many falls inside the cave, this maybe is not even enough.'

Two thousand metres of rope not enough?! Wow, this was quite some undertaking.

'Once we're inside the cave, how wet is it going to be?' cameraman Keith wants to know.

'Think standing under t'shower. In a sauna. In a storm.' Moose is typically understated in his summation.

'That's not good,' muses Keith. 'That means I'm going to have to shoot the whole thing from inside a splashbag.' Splashbags are basically thick plastic bags that you put the camera inside to waterproof it. They're fabulously awkward to film through, and condensation quickly builds up inside. Cameramen despise them.

This comment is met with groans from the two Johnnys. 'No choice lads, it's better to have a poor picture than all cameras dead and no picture at all.'

'First thing we need to get up and running is a hotbox,' Johnny Young decides. On previous expeditions we've just made a plywood box and run a lightbulb inside connected up to the generator. The heat it gives off is enough to gently drive the humidity out of the cameras if left overnight – well, unless a clueless team member decides to put their boots inside to dry out.

'That means we'll have to run the genny quite a lot – will we have enough fuel for that?' Jonny Keeling wants to know.

'It's tight, maybe a couple of hours a night,' Dave Gill chips in.

'How much rope work is there going to be to get down to the cave, d'you reckon?' is my next query.

'Depends how quick ya can go, youth,' Moose responds, as if suggesting it will take me a very long time indeed. 'I'm guessing it'll take at least two hours to get the whole crew down every day.'

More groans and worried looks. That's a heck of a lot of faffing and waiting about before you can even start filming. Then the same back at the end of the day, when you're already exhausted.

'Is there anywhere near the cave entrance we could think about sleeping? Maybe string up some hammocks?'

This is Tim's suggestion. It turns my stomach – the idea of living inside the cave makes me feel genuinely sick. Tim, though, is brought to life by the world's deep, dark places. I remember his face when we were pushing through into hitherto undiscovered passageways in Borneo – his eyes were alight like Gollum in *Lord of the Rings*, utterly electrified by the gloom and the wonders that might lie beneath.

'Forget it, Timbo' – Moose again – 'there's nowhere dry enough to even think of a bivvi, not till you get eight hours-plus into Mageni. It'd just be miserable. Not gonna happen.'

Inwardly I breathe a sigh of relief.

'OK, so the next thing is our evacuation plan,' Tim states.

Moose almost snorts. 'Simple, there isn't one, so don't hurt yourself.'

Johnny Young's ears prick up at this. After all, he's in charge, and if anything goes horribly wrong, he's going to be the one carrying the can. 'Errrr, what do you mean, forget it? I mean presumably we get someone back to the helipad and fly them out to a hospital in Moresby?'

'If there is a helicopter ready' – Jean-Paul shrugs again – 'and the weather lets him here – sometimes one week, two weeks, you cannot land in here. And even then it is two days to get to Moresby.'

'And that's if the problem is here in basecamp.' Moose hammers home his point: 'You boys ever carried anyone on a stretcher?'

Several of us nod. I've only done it once for real, but one memorable eight-mile training exercise carrying a prone colleague will stick with me for eternity. He seemed to get heavier with every step – even with eight burly lads carrying him, it near broke us all.

'It's a two-hour walk from here to Ora village through the jungle,' Moose muses. 'To carry someone on a stretcher ... that'd take maybe five, six hours. Underground, with rope work, through the river, unstable ground, in the dark, one hour would take *all of us* a big day to carry through. And I'm talking a BIG, twelve-, fourteen-hour day, everyone totally fooked and on the edge the whole time. That's for one hour of caving.'

'Then even these ropes outside the cave, she is a day to haul someone through,' Jean-Paul adds.

'Specially if it's a sturdy chap like one of you two' – Moose gestures to Jonny K and me, then looks at Willi – 'and you can forget it for that big lump!'

Willi smiles, and shrugs apologetically.

'Yup, if someone goes down we better pray it's me!' Pam jests. Fair comment. She probably weighs as much as one of Willi's mighty arms. (Tough as nails, though – on one caving expedition we did together she broke her hand and just kept on going!)

'So if you're ten hours inside the cave and you break a leg . . .' Jonny Keeling trails off, and we all extrapolate.

'Once you leave this camp, you are not allowed to break *anything*,' Moose says with finality. 'That way, my friend, lies a world of hurt.'

9

Kicking out from the vertical cliff face, the ropes tangling and twirling beneath me, I swung into open air and with gut-plunging clarity my precarious position was suddenly revealed. Free from the confines of the forest that strives to conceal you within its twisted green grasp, I could at last see that I was in a spectacular vertiginous canyon, like a hairpin in profile. From above, no doubt, it would appear like a giant gouge through the forest, its walls dropping perhaps three hundred metres down to a raging white-water river. Right now, that river was so far below my swinging feet I could have been soaring above it in a passenger jet. But instead I was dangling like a garden spider on a silken thread, my ropes the only thing keeping me from the abyss. Most remarkable, though, was the fact that I was hovering directly above a tumultuous waterfall thundering out of the cliffs, a churning, raging white river dashing down to the rocks far below. The cave that gave birth to that waterfall was my destination.

We discovered the relevance of our safety seminar as we began the process of getting down to the cave for the first time. It was just a short walk from camp to the edge of the ridge that dropped down to the river, way, way below. Mageni cave was about two-thirds of the way up the cliff face.

To safely descend the complex network of ropes, we need to be clad from head to toe in kit. First comes a sweaty waterproof caving suit, which doesn't really keep the water out as we're totally immersed almost from minute one, but does spare you from scrapes, scratches and abrasions dealt out by the fiercely sharp rock. Imagine going into a steam room wearing a full Gore-Tex oversuit, and you'll have some idea of how unpleasant it is. Over the top of this goes a full sit and chest harness with jangling metal technical climbing tools, which rubs, chafes and restrains while snagging on every bit of pointy rock around. By the end of the first day, it is already unbearable. Then there's a helmet fitted up with mini-cameras and with wires dangling everywhere, so I can never take the damn thing off. This is hard-wired into a backpack that takes essentials like first-aid kit and spare climbing gear, and snags even worse than the harness. In order to ascend the rope, you place metal ascenders on to it; their teeth grip the line, you slide them up the rope, then haul your own weight up, as if doing constant chin-ups. In the desperate humidity, the effort and equipment make every second a misery.

The Nakanai Mountains of New Britain are essentially a limestone karst landscape. This is a very distinctive kind of topography, defined by the erosion of limestone bedrocks. Rainwater that's been made acidic from carbon dioxide in the atmosphere and the soil dissolves the limestone. Water flows straight down into the limestone until it hits firmer bedding planes, at which point it starts to flow sideways. (Because of this, karst topography typically has little or no surface water.) Fissures and fault lines are fizzled and eroded, and when the water flows through the rock, in the end it forms tunnels. When these tunnels are large enough for a human being to enter, they become known as caves. You do occasionally get caves in other rock types, but overwhelmingly, because of the chemical action, it is limestone karst that offers up the most to the speleologist.

The first few days we spent getting down to the cave, rigging the ropes and hoiking ourselves up and down them to reach the cave mouth. It was a visually staggering place; the ropes took us down through climber-unfriendly vegetation to the Mageni cave. Here, we were able to genuinely appreciate the cave mouth for the first time. It is a tennis court across and a squash court high, its walls adorned with wispy ferns, palms and ancient dusty spiders' webs. Glossy satin swiftlets bank and weave in the air in front of it, and at dusk hundreds of huge fruit-bats and tiny micro-bats pour out into the misty gorge to feed. And above all of this is the water. It's as if a turbulent river much like the Dart, as it flows across Dartmoor, had suddenly run out of riverbed and just pitched off into space. Your eye is irresistibly drawn to the white spume as it dives over the drop and tumbles a hundred metres down to the deeply blue plunge pool below. It is utterly hypnotic.

Unfortunately, dramatic as our first encounters with Mageni were, we couldn't film them. Already our cameras and sound kit were starting to die in the humidity. We were all set up to film a discovery sequence, where I'd swing round on the ropes and see the cave for the first time. In order to do this, I descended right down to the side of the cave, then hung in my harness waiting for a camera to be ready and the cameraman to get in position. For over an hour, I forced myself to look the other way from the cave so it would still be a surprise, but the cameras just would not come to life. Eventually it got so uncomfortable that cameraman Keith and I had to unstrap ourselves and go inside the cave, to sit and wait for things to get sorted.

We hung around for two hours, getting cold and dripped on, and when a working camera finally got to us my teeth were chattering. Worse, I had such chronic nappy rash that I wanted to beat my head against a wall to stop the itching (I'll never moan about babies crying again). Furthermore, by now I was thoroughly bored with the place. It was one of the only times on an expedition film that I've really had to *act* enthusiastic, when all I've wanted to do is go home and have a nice cup of tea. I'm not much of an actor, and it took five or six takes for me to conjure up anything approaching a believable response, by which time it was getting dark and we had to head straight back up the ropes to camp.

The next day we returned. We were about to broach the dark of the cave for the first time. The bright sunshine of the gorge abruptly disappeared, we switched on our headtorches and waded into the inky interior of Mageni. We waded down a passage that was about the height and breadth of a railway tunnel and the length of a football field, up to our knees in rushing water. Then the ceiling dropped down and we had to stoop with our chins on the water, glass-clear under our unnatural light. Here the passage split, and we chose the right branch, under Moose's expert eye, and clambered up the side of a thundering waterfall. When the bigger lights were turned on to illuminate some shots of us walking through, we became aware that we were in a monstrous cavern, with three other waterfalls tumbling in from the opposite side. The main fall was bigger than anything we have in the UK, the clamber up alongside it rather fun. Over the following weeks we'd be doing this same climb forty-odd times and get thoroughly sick of it, but for now it was an exhilarating plunge into the unknown.

The rock around us was painted yellow and red by our beams. There were very few cave features because this cave is so high-energy; the water flowing through it with constant force. When water with calcite dissolved in it drips from a cave roof, leaving grains of the mineral behind, stalagmites and stalactites are formed. These build up over thousands or tens of thousands of years, the stalactites being the ones on the roof, and beneath them, where the drips hit the cave floor, you find the stalagmites. Here, however, the force of the flow meant nothing had an opportunity to even begin forming. Surprisingly, though, the rock around us and under our feet was still remarkably sharp – I would have expected it to be worn smooth by erosion. My guess is that the power of the river running through Mageni was continually smashing the rocks apart, creating rockfalls and ripping boulders from the ceilings and walls. All the rock around us looked newly hewn, and would be very short-lived.

We pushed on through about an hour of obstacles, before we got to the first section of serious rope work. The tunnel opened out again into a cavern as big as a cinema auditorium; one spindly waterfall tumbled down near its centre. Standing almost at the base of the falls and looking up, you could see there were caverns above us. To access them we'd need to ascend vertical ropes – we'd be swinging around, drenched in the spray from the falls. At the top of these ropes was an awkward sideways traverse with several 'changeovers', where your kit had to be skirted around obstacles on the cliff. Always below you was a massive drop into the black, reminding you that you had to get it right – mistakes would be terminal.

From there, a short ascent via some huge overhanging flowstone chandeliers. Here the cave had obviously been protected by the main flow of the river, and we saw some wondrous examples of natural cave artwork, almost as if a chocolate fountain had been frozen mid-eruption. Another short traverse, and then a weird, slightly saggy ascent, with the rope not held fully taut. This one was a little more difficult, until you got your technique thoroughly sorted. By day three we were all flying through it, but first time around, it involved some pretty hefty bicep work. Of course, we didn't have time to do all of this *and* film it – we merely filmed the first section and then treated the rest as a recce. Over the following days, we'd come back and do each of these tiny sections again, but positioning lights and cameraman first, so that they could be seen by the camera.

Pretty frustrating, when all you want to do is push on into the undiscovered realm beyond.

The next week was testing, to say the least. Five of the team went down with the squits and enthusiastic vomiting, and most were hammock-ridden for at least a day or two. Every morning the healthy would gear up and slog down to the cave, keen and expectant about the hours of exploring that lay ahead. Every day, we'd end up sat in the dark, dank cave entrance for several hours, waiting for everyone else to get down, only to find the cameras fogged and the sound unusable. Keith and the two Johnnys, to their eternal credit, kept upbeat, and they worked tirelessly to drive the ghosts from the machines.

Johnny Young and I, though, really don't see eye-to-eye when it comes to our ideas of what an expedition should be. I want everything to be pared back to its bare bones: the smallest teams, the least equipment possible, in the vain hope of a sort of expedition purism. Johnny's ideal, on the other hand, would be if you could bring an entire working TV studio complete with jibs, dollies and machines that go 'ping' into the jungle. But despite his obsession with technology, and – to me – unfathomable desire to drag several tonnes of useless electronica into the deep bush, you've got to hand it to the guy. He's an old-school documentary film-maker, a superb storyteller, director and media man through and through. His edited programmes are inevitably streets ahead of everyone else's, and he's a major cog in the engine of any expedition. However, even after our three leviathan filmed expeditions, he's still about as comfortable in the jungle as a dad in a discotheque. If there's crotch rot or chiggers around, Johnny will get them. If there are slimy logs to be slipped on, Johnny will find them. But he toils on, doing longer hours than anyone, endlessly trying to coax all his precious technology back to life.

As he's not as experienced with the rope work as Jonny Keeling, it was decided that the latter should do the stuff inside the cave, while Johnny Young should spend most of his days up in basecamp with Dave Gill, battling the swelter by wearing nothing but lycra cycling shorts and a big pair of croc shoes. However much I loathe the cave, I would rate it any time over this life sentence Johnny Young has volunteered for. Because he spends so much more time than us with the few locals who have remained around camp to help, they get to know him better than us, and he acquires a nickname – 'soso-bla-man'. Johnny is understandably chuffed; to be awarded a nickname means they obviously have respect for him and his position as boss of the crew. He is slightly less chuffed when he finds out that 'soso-bla-man' translates as 'man who has breasts'.

It was about this time, while we were tucking into rice and cold Ox & Palm bully beef (our daily dinner), that Dave Gill, who had sold Mageni to the expedition team as a great destination in the first place – and himself as the chief expert on the cave system – let slip that he had never actually been to Mageni before. It had long been clear that he wasn't actually that keen on caving any more, and hadn't done any rope work for years. The Johnnys and I looked at each other aghast, but there was nothing to be gained by having a good old rant – all we really had left now was team spirit and morale.

With nothing to occupy my mind, I started to lose motivation, becoming thoroughly negative about caves and everything in them. I could see Tim starting to eye me with disappointment – I was taking every opportunity to ask how much longer we had left, and moaning about how much I hated big holes in the ground. But even Tim, Gollum himself, the man who usually waxes lyrical about caves like some subterranean evangelist, was starting to look at Mageni as a malevolent presence.

Basecamp offered no kind of relief, either. As we were in a karst landscape where all the water drains straight down into the rock beneath, there was no river, stream or pond to get water from, and we had to rely on collecting the rainwater that ran off the tarpaulins over our sleeping area. As the tarps were pretty grubby, the water needed to be strained and boiled to get rid of all the tiny creatures living in it. Within days, frogs, mosquitoes and other bugs had started laying their eggs in the water buckets, taking advantage of the precious standing water. This, however, was no problem at all compared to when the rains actually stopped for a few days. All of a sudden, we had no option but to haul jerry-cans of water up from the cave, which took hours of sweat and effort.

At night the forest around our tents resounded with a whole orchestra of frogs and insects. Popping out to go to the toilet one evening, I found a huge cane toad, fat and happy in the undergrowth near camp. How and why he had made his way up there is beyond me. These monstrous creatures – I've seen them the size of a football – are natives of Central America. In the 1930s they were introduced to areas of Australasia to combat the spread of sugar-cane beetles. What nobody seemed to have thought of is that the beetles feast high up on the cane, while the hefty cane toads (the heaviest recorded weighed 2.65 kilos, the same as a human baby!) do not climb. Unsurprisingly, their introduction had no effect on the beetles, but the toads themselves spread like wildfire, creating devastation wherever they appeared.

They can lay tens of thousands of eggs (in strings) at a time, and they secrete a virulent toxin from glands behind their eyes that kills just about any native animal that tries to eat them. In addition, cane toads will eat anything, living or dead, that they can get in their mouths. It's Australia that has probably suffered more than any other place on earth from the release of non-native animals: rabbits and even foxes were both brought on ships from the UK, so that early colonists could keep riding out to hunt! These two animals have bred like . . . well, like rabbits, and wreaked eco-Armageddon on Oz. But perhaps cane toads have been an even worse pest. Here in New Guinea a less negative effect has been recorded. But still, to find one here hundreds of miles from any large-scale agriculture I found deeply unsettling.

There were, however, wild delights as well as horrors. At night, our lightbulbs enticed moths, probably a hundred different species, and most of them unknown to science. But without a good key (which doesn't exist) there was no way of knowing which to be excited about. None were larger than our common British moths, but when you examined them up close they were exquisitely pretty. Some were black and polka-dotted with red and peach, others greeny mother of pearl, some scorched-looking and peppered. The most exciting beastie, though, was a huge rhino beetle which bounced about our lights, its wings making a noise like a B52 bomber, threatening to pop the bulbs, seemingly too heavy to keep itself airborne,. He would alight gratefully on any available surface, so I stood below him and he'd bounce between my hands and my head like a remote-controlled toy helicopter.

On their 2006 expedition, the cavers had seen a colony of fruit-bats in a tunnel round the corner from the cave entrance, so we deliberately steered clear of that spot so as not to disturb them before we were ready to film. On the third day we went in, and unbelievably they were still there! Out swooped these huge flying foxes the size of eagles, right into our faces, revealing every bone in their membranous wings, chattering and squeaking in alarm as we moved amongst them. Others dangled languidly from a single toe off the stalactites and imperfections in the ceiling. They made an absolute picture, gently rotating as they swooped around, totally oblivious to us. Then when they swung round and spotted us, they'd do a comical double take, before panicking and flying straight into our faces.

The shots would be quite stunning, we were sure. That evening we returned to camp in high spirits. Johnny had a screen set up so we could replay the footage and revel in our glory, giving ourselves a much-needed morale boost. As the playback began, though, our hearts dropped. Both sound and pictures brought to mind a 1920s Charlie Chaplin film, juddering and jumping and utterly unusable. It seemed that something had gone wrong with the camera. Grimly positive, we returned the next day to reshoot the sequence. Every single bat was gone. They never returned.

Cave environments are a mixed blessing for biologists. They provide constancy in temperature and (flash floodings aside) shelter, which many animals take advantage of. Generally the first hundred metres or so contain the most life, as the cave ecosystem continues to be affected by the outside environment. Here you find many species that head outside to forage at night, such as bats, as well as cave swiftlets that exit during the day. There are also creatures such as moths, and again bats, that may hibernate underground. These animals whose life cycle is only partly bound to the darkness are classed as trogloxenes. In addition, there are nocturnal creatures that use the darkness of the cave to good effect but also live elsewhere, and therefore haven't evolved any new physical characteristics. These include cave spiders, snakes that feed on the bats, crickets and creepy-looking whip spiders with their tickling, tapping, extended front legs that function much like antennae do in insects. Such creatures, which like caves but are not dependent on them, are known as troglophiles.

There are also animals that get swept into caves by accident. I've found bemused, blinking frogs deep inside, and trout that may well have been swept in as fry. True troglobites, on the other hand, live their whole lives in cave systems, and are highly adapted to that environment. Crabs, shrimps and salamanders may lose all their pigment and the use of their eyes, and instead develop long tactile antennae and legs to tap around in the darkness, as well as having increased sensitivity to air pressure and temperature. Their metabolic rates plummet as they find so little food there, and some barely move at all. This leads to long lifespans, with some cave crayfish recorded as living over a hundred years. Such animals can be found quite deep inside caves and have one of the highest rates of endemism of any group of animals – that is, they are likely to only occur in that cave system and nowhere else on earth. Mageni cave had never been scoped for its wildlife secrets before – indeed, no cave system in New Britain had been explored with this in mind. Any true troglobites would have evolved there in solitude, and would certainly be new to science.

It was the seventh day in camp before things started to change for the better. We'd pushed upriver into the cave, climbing huge underground waterfalls, roaring inside their caverns like revving jet planes. Any one of these falls, had they been on the outside and less remote, would have been a sight people would travel across continents to see. However, here in this dark place, only six people had seen them before us.

Jean-Paul and Willi, who had been in that original team, had now pushed ahead to rig the ropes, so we were ascending vertically alongside the falls. To the layman, this probably looks terrifying, but ironically it is easily the safest part of travelling through such caves. The dangerous bit is just moving through the streamway, where the rocks beneath your feet shift and wobble, brittle rock handholds snap under your fingers, and all the time you're walking in a white-water river.

The biggest danger is a slip or tumble, which can break bones, tear ligaments, impale you on the spear-sharp rocks, or pitch you over a high drop to a nasty end below. It's impossible to overestimate what a disaster even a small injury can be. Apart from the ever lurking threat of circumstance, caving is like water torture – no extreme effort, pain or exertion, just the drip, drip, drip of always being wet and mouldy. Plus the darkness, the rubbing and chafing of restrictive kit, the wearing noise of eternally thundering water in an enclosed space, the boredom of moving slowly and cautiously through the same terrain over and over again. It's like death to the scrotum by a hundred thousand papercuts.

The big relief came at the end of every day as we left the cave. It wasn't just being reacquainted with the open air (by the time we emerged it was generally dark, anyhow), but also the chance of a bath in a deep pool on the lip of the waterfall, with the river flowing out over the edge into the open. The bath was maybe a bit pointless, as we'd then have to climb up the ropes outside the cave, which would leave us stinking, sweaty and covered in grime again. But all the same, to plunge butt-naked into the gin-clear pool, the water thundering off into infinity alongside us, was utterly invigorating, and gave us back the will to live. We inevitably had some good chuckles there too, half a dozen stark-bollock birthday-suited cavers scrubbing up in the half darkness – the puckered-up faces and the less than manly sound effects as rugged, real men made the plunge into the chilly water will stay with me for ever.

Having said that the ropes are the safest bit, as I ascended after a particularly bracing bath I had a real scare. I was switching from one rope system to another halfway down the outside climb, when without warning there was a popping sound, my rope whipped free, and I swung out over the void. For a second I hung there, heart pounding and fit to burst out of my chest, but not moving a muscle in case I caused something else to pop. When I finally got up the courage to draw myself back into the cliff to check the rigging, I saw that the bolt holding one of my ropes had just sprung clean out of the chalky rock. In good granite, you could hang a minibus off one of these expansion bolts, but the rock here was such poor quality that they could work themselves free at any moment. All of a sudden, the two remaining bolts above me securing my lifeline seemed far from sufficient. I moved with considerably more caution from then on.

The following day, we were well into the cave and things had gone pretty well. Granted, we'd not managed to film anything all day, but Keith, Tim, Moose and I had pushed on up to another huge waterfall to try and figure out how best to ascend it. On the way back down, Moose took a bit of a detour to look for new 'leads'. We all met up at the top of a climb, and Moose returned with a great big smile on his face.

'Guys, you're not going to believe this, I've found a squeeze that leads through to another river – and there's a big wind coming up through it. It's a completely new passage.'

That night in camp the mood was expectant. We had two cameras working, almost everyone was fit and healthy, and the promise of genuine exploration lay ahead. I now felt thoroughly guilty for not having been my usual positive self, and for having taken for granted this magnificent opportunity afforded us.

The customary couple of hours sat in the soggy cave entrance waiting for the crew to assemble were nothing like as painful on the following day. We left daylight behind and pushed on through familiar territory to find Moose's new lead, which was hidden in a huge chamber, magnificently decorated with stalactites and windblown calcite eccentricities. The main river had made quite a detour away from this place, and for the first time we had quiet. It was incredibly eerie after the non-stop barrage upon the senses of the main cave. For a few hours, it was like being in a weird subterranean world, half expecting to see Captain Kirk wander in, chatting to Scotty on his flip phone. In one corner, a hole dropped away between the boulders. As we climbed down, you could clearly hear the rumble of the rushing water below. At the bottom of the chute was a little cubby-hole, one wall of which consisted of a curtain of water. We had to pass through these wet drapes to get into a narrow tunnel, and from there on to a new streamway. Once we'd passed through the cascade, every single step was the first a human being had ever taken. My enthusiasm had well and truly returned.

The new passage was, however, a real step up in commitment from what we'd attempted before. It was about half the diameter of a Tube train tunnel, heading steeply downhill and with a raging river pelting through it. With no dry ground at the side, the only way to proceed was through the main flow, a real strain even for the heavies like myself, Willi and Moose. Although Pam Fogg is one of our most experienced team members, at half my weight she had to turn back, as she'd have been flushed away instantly. We didn't know for sure what lay below – perhaps a waterfall, or a sump where the water disappeared into airless tunnels – whatever, it was without doubt going to be bad. To have lost your feet and been swept away would be to consign your body and your bones to the cave for eternity.

Within minutes it became clear that this was probably too edgy an escapade for a TV programme. I mean, we all want drama and jeopardy, but none of us want to be on the first expedition team to lose a crew member. Safety adviser Tim, who's been on cave rescue teams for decades and pulled corpses out of such situations before, stopped us in our tracks and told us we should think about turning back. When someone with Tim's cool and experience makes a call like that, you have to listen. The cavers were chomping at the bit to explore, and we were all loath to give up on our first great footage, but Tim spelled it out for us in plain English: 'There's nothing I'll be able to do for you if anything goes wrong down there. Lose your footing, and you will die. It's that simple.'

The strongest of us pushed on down the raging cataract, deeper into the guts of the mountain, proceeding with grim caution and Tim's words hanging over every footstep. At one point I stepped from calf-deep rapids into a huge hole, which engulfed me up to the throat. Water hammered into my face, pressing me back into the hole. Moose reached down with one extendable arm, and pulled me up out before the water could pummel me into submission.

Several hundred metres down the tunnel, and we found that Tim's warning about the consequences of a slip had been utterly correct. The river plunged over a waterfall, which then dropped into a slot in the rock. Being sloshed over that would have certainly been terminal.

As we couldn't continue without ropes, we decided to work our way back, surveying as we went, which involved flashing a laser from wall to wall to assess the dimensions of the passage. This done, we returned over the familiar ground in ebullient mood. Finally, we had some film that justified all the discomfort. Having to haul up our bags in the driving rain and darkness outside didn't cool our enthusiasm, and Keith and Johnny didn't even take off their muddy clothes before checking all the footage. The fear and excitement of this latest escapade was palpable, just viewing the raw material on a three-inch monitor. There were cheers, backslaps and whoops all round, particularly for Keith, who had achieved the near-impossible in bringing the darkness to life. That evening, one of our treasured bottles of Irish whiskey was passed around, everyone taking a tiny capful of the precious elixir.

That night, as if trying to dampen our excitement, a cloud of winged termites took shelter from the rain, gathering around the lights of our camp. The laptop became a writhing sea of insects, and even inside our mosquito nets there was no escape.

These eruptions are an occasional torment in many parts of the tropical and subtropical world. Termites are incredibly important insects, their colonies, millions strong, recycling enormous amounts of plant matter and detritus. They are also probably one of the biggest sources of methane (a much more damaging gas as far as global warming is concerned than carbon dioxide), as they eructate vast amounts of it. The colony usually has just one queen, who has a single attendant king who mates with her repeatedly. She then lays millions of eggs, as many as two thousand a day. She is a grotesque sight, with a normal head and thorax and then a huge swollen sack of an abdomen, yellowy white, and in some species the size of my two fists together.

When the colony gets too large, a winged reproductive caste is produced, called 'alates'. These usually take flight shortly after the first big rains following a sustained period of dry weather. I've not seen any literature on why this happens, but it seems logical that it might be because the softened ground will be easier for the termites to bury into and start a new colony. Or perhaps it's merely that termites from various colonies need to respond to the same signal (the rain) in order to emerge at the same time and commune, expanding the gene pool. Whatever, the alates were taking full advantage of their opportunity to get out and find a mate before starting a new colony. (Somewhat revoltingly, when they land they drop their wings and can wriggle through the smallest holes – which include mosquito nets.)

That night, there was simply no escape from them – we were covered in termites even in our beds. They thronged my face, crawled into my eyes and ears and into my clothing. I know as a naturalist I should be enthused by this display of eusocial invertebrate behaviour, but as a person, I just wanted to get out a can of Raid and blitz the lot. Termites are, though, some of the most fascinating creatures on the planet. Some species include soldiers with huge heads like brown tic tacs and with powerful, mammoth jaws to bite through any attacker. Some can fire an acidic toxin from holes in their heads, others spontaneously split themselves in two, secreting a kind of glue as they sacrifice themselves, which entangles any attacker!

When we woke the next morning, the entire camp was covered in a shimmering mist consisting of millions of discarded diaphanous termite wings. As we went about our business, the draught of our passing made them drift into the air like wind-borne snowflakes.

The termites hadn't beaten us, but the damp did. For the best part of a week none of us had had dry feet for a waking minute, and after a while that takes its toll. Tim and I, the two Johnnys, and worst of all Keith, all went down with trench foot.

Trench or immersion foot is caused by the feet getting constantly soaked, and is named after the Tommies' affliction in the trenches of the First World War. When your feet are wet for long periods and never given time to dry, the skin first wrinkles as if you've been in the bath too long, then it actually seems to rot, turning red and blotchy, then sloughs off from the weeping grazes. It itches like hellfire, then blisters form and burst with rubbing. This may sound like a mere annoyance to anyone reading about it back home in England, but out there, where it's impossible to get your feet dry for even a minute, it is crippling and agonising. Worse, if untreated, the feet swell, serious fungal infections followed by tropical ulcers set in, and necrosis (cell death) starts to take place. In the First World War this led to gangrene and amputations. In New Guinea it just meant we had to stop dead until it was fixed.

It takes an awful lot to take Tim and Keith down, but the day after the first exploration of the new passage they were both hammock-ridden. Once everyone had cleaned their feet and covered them in talc and anti-fungal potions, to put them back in the mud would have required too much hassle, so we just lay all day long sweltering and sticky in our hammocks. No one had wanted to make space for a book in their bags, no one felt the inclination to chat or work on kit. We just lay there, like a bunch of casualties in a field hospital. Later on, we all tried to phone home on the satellite phones. We'd cut down a rotten tree in order to create a small hole in the canopy and hopefully provide a signal for an emergency phone option. Unfortunately, this had only succeeded in producing an occasional three-minute window when the satellite passed overhead. All day long, from each of my colleagues I could hear the exact same attempt at communication: 'Hello? Hello? Can you hear me? Yes, fine, how are you? I said fine. Fine! Hello? Hello? AAAAARRRGGGHHHH!'

That evening, though, was the first time on the trip we could actually say we'd had fun. The bottle of whiskey came back out of Tim's rucksack, and I took a couple of hours to manufacture the best grub possible from our limited supplies. I shallow-fried some local sweet potatoes in the spices from our Pot Noodle sachets, boiled veggies that we'd managed to buy from the folk at Ora, and opened some baked beans which I mixed with processed cheese, Tabasco and peanut butter. It may sound unspeakably grim, and would mean a couple of lean days later in the trip, but after nine days of bully beef and rice it tasted better than my mum's roast Sunday lunch and was a huge boost to morale.

While we sat chatting, several gigantic longhorn beetles flew into the tent. Their bodies were a hand-span long and their long horns (really antennae) must have measured all of five inches. The giant sickle-shaped mandibles looked as if they could have chomped through a lead pencil. I handled them with extreme care. Every few minutes they'd flip their wing casings forward as if popping the hood on a 1950s convertible, revealing their diaphanous wings. Then they'd hover up into the air and buzz in front of my face, making a noise like a box of tintacks being shaken, bobbing around and struggling to keep their huge bulk aloft.

The real treat, though, was that the bugs around our lights were in turn attracting horseshoe bats. The clever winged wizards had worked out that our illuminations had suddenly become the focus for every flying insect for miles around. It was the highlight of our evenings, watching the bats ducking and weaving about and between our mosquito-net strings, flitting past our ears, snatching moths and midges out of the air in dazzling aeronautic displays. Our evening

10

Water surged around my feet – two ships struggling to make it upstream against a turned tide. Just metres in front, the water tumbled out of the cave, free-falling into space. Fractured by tumultuous air currents, the dancing, plummeting waterdrops glinted and sparkled in the sunshine, not seen for days. The spectacular lime flowstone sculptures on the cliff face were hung with orchids, mosses, vines and dripping ferns. Far beneath at the azure plunge pool, winds were blowing fit to crack their cheeks, constant raging cataracts and hurricanes. Down there the water hurtled out into the gorge.

Day nine was the worst and most frustrating day so far of our expedition in the Nakanai Mountains. Our next challenge was the biggest waterfall in the cave, an obstacle that Jean-Paul described as the cave's crux – it had very nearly brought to a halt their exploratory expedition of 2006.

They named the fall the Derbyshire Downfall. It's about ten metres high and five metres across, and the whole force of the Mageni River hammers through it, plunging down into the most immense chamber we'd seen so far. The wind roars through it like a Foehn wind coming down off a glacier, and even on the far side of the chamber it's like standing under a freezing power-shower. It's a mightily impressive and intimidating place, and as much as being a crux of the cave itself it would be a crux point in our film.

When we first reached the fall, we noticed that the earlier expedition's rope was still hanging alongside it. Here, finally, was an opportunity for me to add some of my own expertise to the party. Normally, the team would do what's called 'aid-climbing' to get up the rock beside the fall. This involves drilling holes, whacking in bolts, then the climbers haul themselves up on them – which takes for ever. My alternative suggestion was to save time by solo-climbing the thing, using the old rope as a tenuous backup. I could see a potential line running under the water to the right, and knew it would make an exhilarating and challenging climb.

But Tim, as safety adviser, was dead against it. The rope had been dangling there, constantly battered against the sharp rocks, for over two years, and there was no way of knowing what state it would be in. It was certainly very unlikely to take my weight in the event of a fall. A couple of hours underground, already in one of the most remote areas on the planet, an accident couldn't be risked. If I came off the climb and the rope didn't hold, I'd drop into a maelstrom that I might never be spat out of. Alternatively, the rope might hold, pulling me under the fall itself – and no one would be able to rescue me. Tim, not surprisingly, was very much against the idea. But I pushed my case and at about one in the afternoon – after three hours sat in the dark being dripped on, waiting for everyone else to get into the cave – Tim, Keith and I battled our way around the pool beneath the downfall, and I hauled myself out of the water, teeth chattering and fingers clunky with cold.

From the moment my hands took a hold of the rock, it was clear Tim had been right. It was extremely loose and would come off in my hands. Every time I looked up for a handhold, water would pelt into my face, blinding me, and my trainers slipped about on the wet rock like roller-skates on an ice rink. As a further hindrance, I was wearing a lifejacket and a full rucksack, and the electrical cables from my helmet cam and microphones – running out of the helmet and into the pack – tugged at my head with every move. Nevertheless, with my heart in my mouth I pushed on, forcing myself to concentrate and not to plunge and prove Tim right. On loose rock the important thing is to spread your weight as much as possible, never committing too much of it to one hold. You tug at each handhold and stamp on each foothold before you'll trust it.

Under the pelting water this was all but impossible. The last couple of moves were overhanging, and taking the full power of the water. I tugged upwards against the flow, the cold water blasting in my face. A foothold gave way, bounced once and disappeared into the spray below. I felt my strength going. The backpack was wedged on a big flake behind me – this was it – I was going to fall from the very top and it was going to be *all* bad. I slipped and slithered, panting with fear, but then suddenly a hold I'd not felt before – over the lip of the fall, carved in the shape of an old-fashioned telephone receiver, a big, beautiful Godsend of a hold. Just as my strength was almost gone, I took hold with both hands and bullied myself up and over the top into the cave beyond. Hauling up a new rope behind me, I fixed it in place, ready for the rest of the crew to follow, and looked down to the bank of lights on the other side of the pool, adrenalin zinging, triumphant. I took the old rope in my hands and fed it through. In places, two years of water had whittled it down to a few pieces of burred thread. It wouldn't have taken the weight of a twelve-year-old.

Down below, the lights of my fellow cavers looked like car headlights at a drive-in movie. They were all gesticulating wildly at me, making cross and stop signs, clearly wanting to indicate something was wrong. There was no point in shouting – it would have been like trying to get yourself heard during a launch at Cape Canaveral. What? Had I fixed the rope wrong? I checked everything, but it was fine. Now they were making signs that I should come down. OK, maybe they wanted to regroup before moving on, or get some close-up shots. I abseiled down the new, solid, rope, and swam back across the chilly pool to the crew. Instead of being greeted as a conquering hero, it was all grim faces.

'The camera's fucked,' Keith told me, 'we're aborting.'

The chill was now coming from more than just the water.

'When did the camera go down, Keith? You got the climb right?' I asked with a certain amount of desperation.

Keith shook his head grimly. 'No mate, we didn't get any of that. We'll have to redo it tomorrow.'

There was nothing more to say. For the five hours it took to get back to camp, I just yomped on alone in the darkness. On arriving, I got straight into my hammock without even a word to Dave or Johnny, closed my eyes and shut the world away. It would have been very bad form to show the team how upset I was – they were no doubt just as frustrated. I listened bleakly from inside my mossie net as the boys took a roll call of the equipment. We'd arrived with ten cameras, and were down to one small format camera. Our sole method of righting our kit was the hotbox. The crew decided to run the generator all night in order to keep the hotbox working. It was a big call – the noise would stop us sleeping, and we couldn't really spare the fuel. We did it anyway. Next morning, the boys went into the box to see the results. The lightbulb had blown sometime in the night. It had had no effect whatsoever.

Eleven days in, still over a week to go, and it was all becoming impossible. You didn't have to be especially astute to sense that everyone just wanted to go home. Badly. Unfortunately, things were to get even worse before they got any better.

One of our objectives was to abseil right down to the plunge pool, to see if I could find any animals living there. Keith, Tim and I went down as far as the cave's mouth before trying to rig up the extra hundred metres or so of vertical drop to the glorious pool below. However, as we started to rig, Tim began to look concerned, shaking his head and talking to himself as he fed the lines through. This was ominous. When unflappable Tim starts talking to himself, there is undoubtedly something wrong. 'You're not going to believe this, Steve – I'm really sorry, mate, there's not enough rope here, we're going to be at least ten metres from the ground.'

'Right. Is there any we can take from inside the cave?'

'Nope, the nearest ropes are a couple of hours inside, and we still need them. Sorry, Steve.'

I nodded miserably. 'This just isn't happening is it? Come on, let's jack it in.'

I could see the relief in both Keith and Tim, and we wasted no time in getting back on to the ropes. With every single jug, I cursed New Guinea and everything in it. This just seemed to be a place where I was destined to fail. Perhaps I should have learnt from my earlier trips and not come back.

As we walked, disconsolate, into basecamp, the two Johnnys were sat in front of a viewing monitor and barely even looked up. They didn't ask us what we were doing back so soon, which would usually have been their first concern. They were anxiously going back through the rushes (the raw film before it's edited). It seemed the camera fault that'd destroyed our bat sequence was present in *everything* we'd filmed on the big high-definition camera so far. Everything, right through from the dolphins outside the Cocky Lady's Guest House, through arriving at Ora village, discovering the cave, getting inside for the first time. We'd come here to make this film, and these were essential ingredients, so they would have to be done again. My New Guinea curse was striking again.

That evening was unspeakably bleak. The two Johnnys worked out which stuff would have to be – soul-destroyingly – consigned to the bin, or redone. From my point of view, too, it was a nightmare. Sequences in which things had been at least close to real the first time round would now have to be done two weeks after the fact. They would look and sound fake. Neither the viewers nor the people who hired me would care about why that was, they'd just write me off as being insincere or lacking enthusiasm.

The next day Willi and Jean-Paul left at dawn to go deep into the cave and retrieve some rope so we could again try to film the plunge pool, this time taking all five working cameras with us. Tim and the Frenchman slogged their guts out to get the whole thing rigged for late morning, so Keith and I could go down and try and film something.

As soon as I dropped down alongside the waterfall below the cave, it became clear this was a very special place indeed. You couldn't really discern from the entrance quite what an overhanging lip the waterfall flowed over, tumbling out of the cave, then free-falling. As the ropes made contact with the flowstone of the cliff face, they'd clip chunks of vegetation – moss, ferns, orchids, vines – unloosing leaves and blossoms that would slowly spiral downwards, like helicoptering sycamore seeds, to the plunge pool beneath. The ropes extended about ninety metres alongside the main fall of Mageni. Below, I could hear a roaring gale, driven by the falls, blowing the water out into the gorge – I felt like I was like standing on the bow of a ship in the North Sea in a force 10 gale. From the larger plunge pools the water flowed away over several smaller ones, then to the river at the bottom of the gorge, perhaps another sixty or so metres down. Each fall ended in a smaller pool, an emerald grotto surrounded by lush foliage. Looking up, the green cliff towered above, and the three mouths of Mageni issued forth their white curtains of water.

Here was one of the most staggering Lost World locations I've ever seen. It was a punishing, physically draining day, but the shots were awe-inspiring, and this time all seemed to be going brilliantly. Mid-afternoon we reached the main plunge pool, dropped into it up to our knees and waded across to try and find some shelter from the ferocious cloudburst the falls were letting loose on us. Two cavers from the last expedition had ventured down here, but the topography of the place was such that no one had followed in their footsteps. I was the third person ever to see it.

Keith and I found a place behind a giant boulder where we could get a little shelter from the wind and rain, and took stock of what we had in the way of working equipment. Keith took the main camera and went to switch it on. Nothing at all. Totally dead. The second one registered an electrical humidity warning, so wouldn't allow us to record. The third was fogged inside on all of the lenses, the fourth the same. And then the last tiny Handycam, a model known for being average quality but a real workhorse that'd never let you down – even that was showing no signs of life. All five of the cameras were dead.

We just wanted to lie down and cry. We both thrive on adversity, but if the kit and the circumstances won't allow you to do your job, all that resourcefulness and enthusiasm counts for nothing. We looked at the prehistoric paradise around us. We should have been bursting with excitement at the opportunity to explore this unknown world, but could do nothing except dejectedly get back on the ropes and begin the long journey back to basecamp. As we made our way back up, Keith developed a nasty blister on the palm of one hand and had to hitch himself up over several hours on raw, bleeding flesh. Never one to complain, he pushed on regardless through gritted teeth, seeking the sanctuary of camp, such as it was.

Perhaps the worst kick in the teeth came many hours later when we finally limped into camp, exhausted, dejected. The two Johnnys watched the little footage Keith had managed to film, and were highly critical. 'This'll never make it into the final film, there are no animals, no resolution to the story – it looks great but it just doesn't go anywhere.' In all fairness, they were facing their own demons at the time, but it was an almighty blow to our morale, and just served to add to the spiral of disillusionment. Suddenly, our crew of normally unflappable people were getting niggly and gripey, putting their own misery before the feelings of others. The usual bonhomie was replaced by a general short-temperedness, which – while it never actually boiled over into angry words – certainly didn't help matters.

Finally, at the fortnight mark, we got some respite.

We were pushing on back to the end of the new lead we'd followed a few days before, with the spare ropes we'd now got from the plunge pool. The hotbox had been working all night long, and five cameras were up and running again. We spent the customary first couple of hours sitting around inside the cave getting cold, bored and damp again, but it wasn't so desperate now, knowing that finally we had a chance of getting something on camera. As we tramped through the curtain of water leading into the new passage, our hearts started to pound again and all selfish thoughts of suffering vapourised with the spray.

The lead had finished at a short but violent waterfall, but this time we were prepared. We rigged a rope system to abseil down through the fall into a narrow slot, pounded all the time by the rushing water. From there on in, you could only make yourself heard by screaming out your words at those standing right next to you, and even then it was no better than Chinese whispers.

'STEVE ... TURN YOUR HEADTORCH OFF, YOU'RE BURNING OUT ON CAMERA.'

'YOU WHAT? LOOKS LIKE BURNING A CANDLE?'

'NO, YOUR TORCH – *OFF*!'

'YOU WANT ME TO MOVE OFF?'

And so on.

Down below the furthest falls was the edgiest scenario we'd yet encountered. We had to traverse several rapids, then a vertical drop, after which the river disappeared under a rock ceiling that descended to within a foot of the water, which flowed deep and dazzlingly ultramarine. Moose and I swam ahead with a safety line attached to us in case we got swept away, with mighty Willi paying out the rope. The ceiling was no more than a long duck, then the cavern opened out again. Our lights cut light-sabre beams into the river as it gurgled and gushed, and gave the place the feel of a magic grotto. It was a wondrous secret lying far beneath the earth that had never, ever seen light before, and it was *my* eyes that were the first to see it, my feet the first to clamber over its rocks. Trying to film, check the exposure, framing and focus while simultaneously keeping my footing and making sure there was just the right amount of light coming from Moose and my torches was near impossible. How Keith had managed to do it all day and every day was just beyond me, especially now with his hand missing a layer of skin, raw and sore.

Our spell-binding tunnel continued for the length of a football pitch, dropping over one smaller waterfall before the water pooled out and seemed to run straight into a wall. Moose warned me to stay well back from the wall, where the water was being sucked with huge force down into airless passages in what cavers call a sump. Doubtless it would re-emerge somewhere further downstream in another passage, but no one would ever know where. This was our end point, and probably where our film would triumphantly conclude.

For the next few days we did various runs into the caves, redoing all the stuff we'd filmed before but lost due to camera faults. It was a pretty thankless task, but at least we were getting footage, and coming away at the end of each day with at least one camera still working.

One evening, though, after our usual lug up the ropes, Keith was found with a pair of tweezers, plucking bits of grit out of his hand and looking stern.

'This is getting infected.'

'Jesus, mate!' I grabbed a hold of his hefty climber's paw. 'When did it start looking like this?'

This attracted the attention of the others, interest suddenly piqued by this unexpected drama.

'How d'you do that, Keith?' asked Pam.

'It was just a nasty blister, I got it from jumaring up the ropes.' (Jumaring is the process of hitching yourself up vertical ropes.)

'Weren't you wearing gloves?'

'Almost all the time, but I have to take them off to film. I think some grit must have got into it.' His hand was already angry-red and swollen. It didn't look good at all.

'KP, I hate to break this to you,' I said, 'but this is infected already. You need to scrub this cut right out with iodine – I'll sort you out some antibiotics.'

That night, Keith took a scrubbing brush and some antiseptic and scoured the wound out, with gritted teeth and a few stifled yelps of pain. He took the absolute maximum dose of antibiotics, and hit the hammock good and early. However, out here, where you are eternally soggy and riddled with bacteria, there's precious little you can do to counteract an infection once it begins.

The next morning Keith got out of bed looking really grim.

'How is it this morning, KP?' I asked tentatively.

In answer, Keith held his hand up to show me. It looked like an inflated red washing-up glove, with an especially big, firm swelling across the knuckles.

'Not looking good, Stevie B,' he replied. 'I've been up all night long with a fever. Those antibiotics haven't touched it.'

'Holy God, KP, your hand looks like a bunch of bananas!' I put my hand to his forehead. He was still burning up, his body battling against the devils within.

'And I don't know what this is, but it can't be good' – Keith turned his arm over to show the veins. They were red and inflamed all the way up to his elbow, as if something insidious was swimming up through his circulatory system. 'That infection gets into the bones, and I'm in big trouble. A cameraman without his hand isn't much good to anyone.' It was the first and last time I've ever heard Keith sound scared about anything.

It was time for a pre-breakfast war council. We sat out in the open while Johnny Young struggled to get through to the BBC's medical support team on the sat phone.

'Yeah, it's Keith, he's got an infection in his hand. It seems to be spreading up his arm, and he's feeling really rough.' He battled against the reception, the crackles, for a few minutes. 'Evacuated? Yes, we kind of thought so . . . to Brisbane?! Oh God.'

This was pretty hard core. We'd assumed they'd just evacuate him out to Port Moresby, but it seemed the BBC weren't confident enough in the hospitals there and wanted to get him to Australia.

'Works for me,' said Keith. 'It's going to take for ever to get me there though.'

The next call was to the chopper operators to see if they might be able to get in to Ora. Luckily the weather was pretty clear, and the helicopter was mobilised that very day. Keith packed up straight away, and set off for Ora village. There were hugs and good wishes all round. But quite honestly, there wasn't a single one of us who didn't yearn to be on that chopper too, to get out of the mud and into a nice clean hospital bed.

After Keith had gone, there was a real low while we all figured out what on earth we were going to do next. It looked like the two Johnnys would have to take on the filming job, which – as I had just found out – was almost impossible. What was worse, we had already filmed everything that was accessible within a day's reach of camp. Our only way to stand any chance of penetrating right into the cave was to push further on and to stay inside it for several days. We were all dreading that. Doing it without our cameraman was unthinkable.

The general mood was summed up when Jonny Keeling took off his socks to get a look at his itchy trench foot. An odious smell hit us, much like fortnight-old summer dustbins. His socks were crawling with maggots. It was beginning to seem as if the jungle was reclaiming its territory, growing not just around us but on us and in us, infecting us like some insidious disease.

As we were spending all our time either down in the cave or cajoling the cameras in a vain attempt to get them to work, I decided to cut a few corners and try and find some more wildlife. Wesley and Thomas were two young men from Ora village who were staying with us, doing odd bits of carrying and camp chores. I offered them a bounty of two *kina* (about forty pence) for every unusual beastie they could find for me. The next morning, a loud chattering and laughing was heard approaching like a happy freight train, and about thirty children poured into camp bearing assorted minibeasts. They'd walked the two hours from Ora village with hands full of bugs for me. Not only was I over the moon to get some new specimens, but it was also a joy to see some smiley faces after the misery-guts expressions of all my colleagues! The kids ranged from about six to thirteen years old, both boys and girls, with identically short-cropped hair and wearing tatty Western cast-off clothes, but adorned with colourful beads around their necks. The girls also carried woven *bilum* bags down their backs, the straps over their heads like multicoloured headbands.

The skins of some of them were infected with *grile* (ringworm), and all had green snot streaming from their noses – even though scars on their upper arms indicated they'd been given their inoculations, undoubtedly by local missionaries. All the same, we were a little concerned about the snotty noses. We'd been warned that one of the easiest ways to get a serious infection, one that might lead to a tropical ulcer, would be to shake hands with a Papuan who had recently been using that same hand to excavate a nostril or lance a boil. Having seen the horrors of tropical ulcers, we were all loath to shake hands with the kids, but made a quick decision that we'd rather get sick than offend them. They seemed a happy, cheery bunch, though the youngest were clearly scared stiff of us all, and would sprint for cover if we so much as looked at them.

The bugs they brought were a motley assortment, mostly the common bush crickets that had been the main players each evening in our jungle orchestra. There were also, however, some real gems. One huge stick insect looked like the bug version of a Sherman tank. It belonged to a group that have evolved a wonderful way of exploiting their neighbours, by dropping their eggs into the path of kleptomaniac ants. Much like the adult stick insects mimic sticks and leaves, their eggs look exactly like seeds – even, like seeds, having a fleshy structure rich in fats. Ants like to feed these fatty morsels to their young, so collect the eggs and take them back to their nests. There, they cut off the fatty bit and chuck what remains (the actual egg) on to a sort of rubbish tip within the ant nest. The eggs hatch out and the young stick insects arrive in the world under a protective escort of a hundred thousand stinging ants. When they emerge, they're not so different in size and shape from the ants themselves, but after about five or six moults they've become huge great spiky beasts, the size of a TV remote but with vicious barbs over legs, body, head – everywhere, in fact. They look as if they could kill you with a single bite, but are actually harmless leaf-eaters.

The kids had also brought a huge golden orb-weaving spider. They're not especially unusual to find here, but it was incredible that they were so comfortable holding them in their hands. This is a scuttly spider with an abdomen the size of my thumb and legs the length of my forefinger, which would send any kid in the developed world screaming behind the sofa. In fact, probably the most remarkable thing about our bug fest was that all these children were carrying their bugs like professional entomologists would, restraining the crickets at the thorax so they could neither leap nor fly, but allowing the stick insects to wander free so as not to risk losing any of their limbs (as a defence mechanism these creatures will autotomise – deliberately drop a leg – if a predator has a hold of it). The kids had undoubtedly caught all of these things the night before, then brought them in their hands unhurt over a slippery, muddy, two-hour walk.

My favourites of the haul were four or five of the most brightly coloured stick insects I've ever seen. There were two similar species, the females about four times the weight and size of the males (about as big as a blackboard marker) and with remnant stumps of wings, while the males had functional parachuting wings. The base colour was a metallic green, so dark as to be almost black. Then down the back were wide horizontal stripes in black, almost luminous green, and yellow. One species had vibrant orange legs and black and yellow stripes like a bumblebee. These are certainly aposematic – bright and warning – colours, and I'll bet a fortune these sticks produce defensive toxic secretions and would taste like sick! I'd never seen anything like them before, and we took lots of photos so I could try and get an ID after returning home. My inquiries were to draw a blank: they were just one of tens of thousands of species surrounding us that were totally unknown to science.

It was such a pleasure laughing and joking with these delightful little kids, there in the dappled sunlight, that putting on the mouldy wet kit and heading down the ropes was even more unpleasant than usual. Even more so, because this was to be the day when we would head deep into the cave, pushing on into new territory and spending the night underground. We would all be very weighed down, carrying all our overnight gear plus three cameras and assorted lights. This makes cave travel really arduous, but I had the added pain of the helmet camera, which lives in a waterproof case in my rucksack, and, as mentioned before, is attached to my helmet by a bundle of wires. This means I can *never* take either my helmet or my rucksack off.

Until you have to climb vertical ropes, this is a mere annoyance, You would normally drop the bag below you on a short leash, which keeps you balanced. However, I can't do this, and abseiling or ascending with a heavy pack on your back means you're doing permanent sit-ups just to keep upright. If you get tired or lose your grip on the ropes, you tip upside down and just hang there with the blood rushing to your head until you can recover enough energy to continue. It's utterly draining. Add to that the annoyance of struggling to get through low passageways with a pack on your back, and by day two of the trip I was secretly praying the helmet camera would just quietly explode so we could consign it to the dustbin. Unfortunately, it proved to be about the most resilient piece of camera kit we had. At times it seemed we would be shooting the whole series on that one camera, which was the size of a lipstick.

The next chat had to be about how on earth we were going to film the rest of the trip.

'Well, I've put it off long enough, I think it's time I got my caving suit on,' said Johnny Young.

A sharp intake of breath followed from everyone.

Tim had to be the one to dissuade him. 'Johnny, you've really struggled with the ropes even just getting down to the cave. This is serious, serious stuff we're getting into here. Can't JK do it?'

Jonny Keeling is a natural athlete, and looked as comfortable on the ropes as the professional cavers. It was the natural choice. Though there was no doubting that Johnny Young was the most competent and experience cameraman.

'I'm more than happy to do it,' he replied, 'as long as everyone's prepared for the footage to come back a bit ropy!'

'It's just too much to ask of JK,' Johnny Young countered. 'This is a major part of our film, this has to be my responsibility.'

Tim was biting his lip now and was visibly starting to sweat. 'Johnny, with the best will in the world, if you take a tumble deep in the cave then we're in big trouble. And some of those rope ascents inside the cave are very very tough. This could be putting the whole team in danger.'

Johnny nodded. He was taking it all on board, but was still resolute. 'Listen Tim, I totally get that, and at the end of the day it's your call, but we're here to make a film. There's no point you all going three days into the cave if we don't get it on camera. If the audience don't see it, it never happened!'

'Surely if you just stick with Johnny the whole way he'll be all right?' This was Pam. She – like all of us – had total faith in Tim's ability to watch Johnny's back.

Moose agreed: 'Aye, and if he cops it, we'll just wrap him up in lifejackets and float him out!'

Despite all our trepidation, the journey into Mageni that morning was a joy. As our filming objectives were located further inside the cave, the cameras stayed packed away in our bags. We staggered a bit as we entered the cave, just managing a decent pace. We were finally getting to really enjoy the caving without any tedious sitting around.

That day was a total epiphany for me. Tackled at a natural pace, Mageni was a wondrous place, the best fun you could possibly have underground. Granted, it is a very committing cave – much like a sublime canyoning expedition done at night. You wade through raging rapids, scale awe-inspiring waterfalls and skirt sapphire-blue pools, before tiptoeing through silent stalactite-ceilinged cathedrals. Moose led the way, sniffing like an inquisitive spaniel down any unexplored side passages, opening up endless chambers that will probably never again see light.

For one member of the team, though, it was not a happy experience. This is not a judgement on Johnny Young. Mageni may have been a highly achievable objective for people like Willi, Jean-Paul and Moose, but they are some of the most able and experienced cavers in the world. Johnny, by contrast, was operating outside his comfort zone, and not surprisingly Tim was terrified that he would take a slip and the whole team would have to spend days hauling him out.

To begin with, Johnny did brilliantly, getting down into the cave entrance with no problems, managing all of the rope work, and keeping pace with Moose and Willi right up until we reached the Derbyshire Downfall – the big waterfall I had free-climbed the week before. Now, bedecked with conventional ropes, the fall presented a tricky ascent with water pounding in your face and a section that redirected you midway. It was quite a short pitch, and we sailed through it in four or five water-blasted minutes. However, Johnny made an easy but vital mistake at the rebelay (where you re-anchor the rope), moving his climbing devices up way too high and getting stuck. Had one of the rest of us done it, we could probably have yanked ourselves through it with brute force and ignorance, but Johnny was already exhausted from the five hours' worth of ropework and caving that had got him there.

We watched, powerless, as he battled with his ropes, freezing water thundering into his face and pounding him into submission. It was the absolute worst possible place to get stuck, the only solution a massive burst of strength he had simply run out of. After twenty minutes of watching him struggle, Tim had to drop down the ropes to try and rescue him. He took Johnny's bag and managed to fix his ropes, but by then Johnny was so spent that all he could do was hang out in the waterfall, praying for some strength to come back. This was a really dicey moment. If he wasn't going to be able to carry on we would simply have to abort and head for home. Somehow, though, he found the last few shreds of energy, and made it up to us in the pools above. He'd been three-quarters of an hour hanging there in the furore of the falls, and was exhausted. However, with typical bulldog spirit he decided to push on.

We found something up there that defied belief in a hundred different ways. It was a hefty crab, which I caught on the edge of a pool way above the Derbyshire Downfall. It wasn't a cave-adapted crab – it had fully functional eyes (on stalks), large chunky pincers and deep-brown pigment, and the body was about the size of my clenched fist. It looked like something you pull up on your fishing line off the pier at the seaside. The only possible explanation was that it must have been swept in from a sinkhole somewhere above, most probably as a larva, when still tiny enough to percolate down through the limestone.

What was even more remarkable, though, was that the carapace and legs of the crab were crawling with bizarre and repulsive parasites. They were white, covered in mucus and vaguely worm-like, inching across the crab's eyes and in and out of its gills. It turned out that these were branchiobdellids, a parasite known for exploiting decapods (ten-legged beasties) such as crayfish and feeding off the organic matter they find on their hosts' gills. But it's possible that their presence actually aids the survival of the host crab or crayfish by cleaning up the gills at the time. I didn't know any of this then, though, and they just gave me the creeps. I imagined myself like something from *The X Files*, discovering an unknown parasite in the depths of the earth, getting infected and then – I don't know – probably my ears leaking black tar and then my head exploding.

The next kilometre or so was the most exciting, beautiful, *visceral* stretch of cave I've ever been in. We were wading up the course of the main river in a beautiful, wide passage, not stopping to film much so having a whale of a time. The last challenge of the day that we encountered was where the river ran deep and the roof was so low that you had to swim, pulling yourself forward with the aid of the low-hanging stalactites. It was like swimming up the gullet of a giant sea beast, and tugging on its tonsils to help you escape. Finally, after around twelve hours on the go, we found a small passage to the side of the main river. After the roar of the Mageni battering our ears all day, it had an unearthly quiet, accentuated by the staccato plink-plunk of water dripping from the stalactites. The floor was ancient mud, which had probably been deposited there hundreds of thousands of years ago, and although it stuck in cloying lumps to our feet and our suits, it was just right for flattening out patches to sleep on.

The chamber was elegantly hung with giant formations. There were eccentric helictites, where the calcite had gone crazy and started to grow sideways and upwards in weird crystal shapes. They looked like curly fries, worms, doylies, thin brittle stalactites that had grown in zero gravity. In reality, their shape is probably due to winds blowing into the caves as the calcite was deposited, lifting droplets of water off in strange directions. Helictites are impossibly fragile, and despite being thousands of years old would splinter and break off even if you brushed past them ever so lightly. There were also huge stalactites that had fallen off, then had more calcite form on top of them so they looked like the ribcage of a long-dead sea monster or the timbers of a weird subterranean shipwreck.

It was in the back of the cave, though, that we made the most memorable discovery of the trip. There we found a shallow pool that held several hundred tiny shrimp, and maybe twenty or thirty pure-white crabs. The crabs had bodies the size of a fifty-pence piece, yet each leg was as long and thin as the ink tube inside a biro. Their pincers had an extra spike on the main joint (for what purpose I have no idea), their eyes had atrophied to nothing at all, and they didn't react at all to my light. Both of these species had evolved their adaptations to life in the dark right here in Mageni – they could only have been washed down into this cave as eggs or microscopic larvae many millennia ago, and are – without question – new species to science.

The cave floor made for a lumpy, bumpy bed, and despite my fatigue sleep didn't come easily. Whenever I didn't have my mind occupied with some task, a deep sense of unease started to creep in from the shadows and I felt a brief, heart-pounding panic whenever I started thinking about the fact that I was as good as entombed two miles beneath the earth. The only solution was to fill my mind with irrelevant stuff – and a busy mind is the mortal enemy of sleep. This was just the beginning, though, and things were about to get seriously bad.

Eleven at night, and I woke up without a clue where I was, and feeling really rather peculiar. I was freezing cold and just couldn't get comfortable – even wrapping myself up in my sleeping bag in the foetal position was no good. An hour later, and I woke again in a real state this time. From nowhere, I'd developed a full-on fever, veering from raging sweats where I couldn't bear to wear even my lightest clothes, to violent shivers even when tightly encased in sleeping bag, clothes and bivvi bag. Though it didn't spring to mind at the time, it's possible that I was suffering from malarial relapse, which can present symptoms exactly like these for as long as six to ten hours, years after you've suffered from the disease.

It continued unabated all night. I tossed and turned, hallucinating about being buried and trying to scream out but unable to open my mouth. I was about as far away from a hospital or any other kind of rescue as it is possible to be, and honestly felt like I was going to die. If it was malaria, we were all doomed. That powerful, all-encompassing heat inside that totally incapacitates you, a fatigue so intense you end up lying in your own waste as if the bathroom was a million miles away. How on earth would I manage to get out of this place when even the thought of getting out of my sleeping bag was more than I could bear?

Next morning the others started to rouse themselves, and I tried desperately to convey to them my condition.

'Boys, I'm in a shit state, I've got a raging fever, I've been up all night.'

'Mate, we need to push on in, we have to find where the cave goes – that's the ending to the film!'

'Forget going on, Johnny. I'm worried enough about even getting out. Honestly mate, I'm in pieces.'

'OK. Well listen, we'll give you a couple of hours to get yourself together, then we'll reassess.'

Understandably, they were a bit nonplussed. While they milled about sorting themselves out in silence, nobody offered help or showed any care. At first I burned inside with the knowledge that they were angry with me for putting the mockers on the trip. But I was absolutely furious – couldn't they see how ill I was? Later, I realised this wasn't it at all. In fact they could see perfectly well how sick I was: that was the problem. There was nothing they could do, though. I had to get out under my own steam. The silence was just them all silently thinking about how horrific it was going to be to get me out of there if I couldn't rouse myself. There was no point being precious with me. It was a battle I had to fight on my own.

A couple of hours later the fever had abated, but in its place was total exhaustion. All I wanted to do was lie where I was until the demons took me. Just getting my trainers on seemed to require more energy than I had and to take about ten minutes. All I could think about was making things easy for myself. First off I asked if I could leave the helmet camera off for the way out as the rucksack and wires were just too much for me. Jonny Keeling said he'd take the camera, but Johnny Young said no, as it would cause too much of a continuity issue. So in my decrepit, exhausted state they proceeded to rig me up in the dreaded thing. Then, as we were swimming out through the 'Lake District', as someone had dubbed it, Johnny asked me to do a bunch of set-up swim-through shots, treading cold water while swimming through stalactites.

'Steve, mate, can you give us a piece to camera about this place? – it's pretty spectacular.'

I launched into a piece to camera that focused on how bad I was feeling, but Johnny stopped me mid-flow.

Above Sat contemplating the last few rays of sunshine on the small plunge pool at the entrance to Mageni cave.

Right The red suited munchkins get ready for an overnight trip into darkness. Johnny Young is to right of shot, Tim Fogg is facing camera in the centre of the picture.

Pulling myself up out of a puddle as we headed deep into the cave on our overnight trip.

FACING PAGE

Top Wearing a grim expression as we pushed into uncharted darkness deep inside Mageni. Safety guru Tim Fogg had just told us any mistakes here would result in certain death. We were all fiercely focused.

Centre This is me getting on the ropes at the bottom of one of the small falls inside Mageni. We had to navigate endless stretches of vertical climbing to access the uncharted areas of the cave system.

Below John Paul stood in the mouth of Mageni cave, a few metres from the waterfall that plunged eighty metres down to the gorge.

THIS PAGE

Right Mageni cave as it erupts from miles underwater. The central and largest entrance is the one we used to access the cave.

Below & inset That tiny red spider on a string of silk is actually me hanging over the mouth of Mageni, photographed by helicopter.

Me dangling on the ropes over the top of the entrance to Mageni.

Top The king bird of paradise, a glorious crimson miracle down in the jungle darkness.

Above Cameraman Gordon Buchanan admires the pygmy parrot – shortly before it savaged his finger and made a break for freedom!

Above The beautiful fruit dove. And that's not me being descriptive; that's its actual name!

Left The pygmy parrot that savaged big brave Gordon Buchanan!

Below The hooded pitohui, one of the world's only poisonous birds.

Above This angle-headed lizard shows off his dramatic gular flap as he sits waiting to display to a female.

Right The small-eyed snake that wasn't a small-eyed snake!

Below With one of the female phasmids or stick insects brought to me by the charming kids from Ora village. We couldn't take any specimens out of New Britain, but I'd be stunned if this was not a new species.

The Bosavi silky cuscus, an adorable marsupial with a delightful character, and never before seen by science.

George holds a scorpion under ultraviolet light.

'No, no, this needs to be from yesterday, as if you're seeing it for the first time.'

I stopped and looked at him to check if he was serious. This was not something we ever did in the expedition films, honesty being everything. However, I didn't have the strength to argue.

We did it. But Johnny hadn't framed it quite right, so he sent me back to swim through it again. My teeth were chattering, I was fuming at having been made to wear the silly camera, the stupidity of having to pretend this new and horrid day was yesterday. My mind was just totally focused on getting the hell out of the cave.

Looking back, I should have just done it – it would have been over in a minute and we could have gone on our way. Johnny was, after all, just doing his job. But it had all simply built up, all the frustrations of the last few weeks, the curse of New Guinea that still seemed to be infecting me, the anger at being asked to act, which I hate doing, the exhaustion, the fear. I'm normally quite a calm and rational person; I've got through my entire television career without notable tantrums. But at that moment I was beside myself with fury. To my shame, I shouted at Johnny: 'Are you having a laugh? I'm running a raging temperature, and you're asking me to do swim-throughs in freezing cold water? And pretend it was yesterday? You must be fucking joking!'

It's the cardinal sin. You never shout at your team members, particularly not when things are bad. Team unity is what gets you out of things like this. I did apologise later, but not until the nightmare was long over.

To somehow make matters worse, I actually got out of the cave under my own steam. Once I'd put my head down and just single-mindedly went for it, I even outstripped Moose. I barely stopped going the whole way. When I crashed into my hammock later that afternoon, I dropped straight into a coma-like sleep that lasted seventeen hours.

This was pretty much to be our last effort, and it seemed the only real horror now was going to be actually getting out of the place – an effort that would be sweetened by the knowledge that every step was a step closer to home.

The escape, though, was not to be as simple as we had hoped. When we came out of the cave it was to find a row of unfamiliar faces by the top of the ropes. It seemed they were from another village across the gorge. Indeed, they were not just *from* another village – they *were* the village. Every man, woman and child had come across to look at us. However, whereas the people from Ora had treated us with a kindly bewilderment, these new folk had a totally different feel about them. Apart from the kids, they didn't break into easy smiles when you grinned at them, and they walked through our camp going through our stuff with an attitude we weren't used to. It turned out there was a reason for this. They claimed ownership of the land we were on, claimed we'd hacked down a sacred grove of trees where they'd buried their ex-chief, and that we were paying Ora village when the money ought to be going to them.

Throughout my travels in New Guinea, I've had two reactions from people who've never seen white people before: either the friendly inquisitiveness we'd experienced from Ora village, or outright hostility. The latter can turn bad very quickly. After all, these may be people who are not a generation away from wholesale headhunting and cannibalism, and in many cases still settle disputes with brutal aggression. All of the men were brandishing machetes and axes, and were savagely chopping into every tree, branch and plant in sight. The intention was clearly not to open up an area of forest to make a garden or clear a nice sunny spot to enjoy an evening canapé. This was without doubt the equivalent of a baboon yawning extravagantly to show off his huge canines; a seemingly nonchalant gesture, but actually a premeditated measure to show any potential challenger quite how dangerous you might be.

That night we were to be low on sleep, as the hundred or so tribespeople argued, chattered, sang, and did huge comedy yawns and sneezes all right in our faces. The yawns were particularly annoying – yelled practically as loud as it is possible to yell a yawn. I mean, honestly, if you're that tired, go to bloody bed! The noise was ceaseless, and penetrated clear through both earplugs and the pillow I'd rammed over my head.

There is simply no concept of personal space in New Guinea. Take the following example. The hut in Ora where we filmed on the first night is smaller than my kitchen, and thirteen people were sleeping there side by side. People here think nothing of sitting with any and every part of their body touching you, and just staring at whatever you're doing for hours on end, even following you as you wander off for a poo. There is nothing inherently bad in this; after all, isn't it a little bonkers that we in the Western world are so precious as to consider taboo the most basic of bodily functions, things that every one of us has to do just to survive? And isn't it weird that we've (only very recently) acquired this hyperdeveloped sense of personal space?

Social historians suggest that the great value we put on personal space in Western societies dates to only about two hundred years back, to the Romantic poets who for the first time in living memory idealised solitude, nature, and wandering 'lonely as a cloud, that floats on high o'er hills and vales'. Before that, English society saw infinitely more sense in a life crammed next to your neighbours day in day out, in the same way as the Papuans do today. After all the time I've spent in the developing world, and in these kinds of environments, I am a little ashamed that I've not been able to get used to their different perceptions of closeness. Unfortunately my Western sensibilities are just too deeply ingrained, and I quickly start feeling really claustrophobic.

No doubt it was all the more unsettling in this case because there was an underlying current of aggression behind it all, and every time we turned our backs someone would be going through our gear, filling up their *bilum* bags with whatever they fancied. This ranged from our food to head-torches and clothes. When caught in the act, they would stare at you defiantly, then grumpily put the stuff back again. As soon as you turned your back, they'd come straight back and start filling up their bags again. Being as we had nothing whatsoever in the way of surplus with us, this was a real issue, so we couldn't let our guard down for a second.

The next morning as we started to pack up our gear for our final escape, it began to seem as if it was all going to go badly. Porters from Ora village were turning up to carry our stuff for us, so that two camps were forming, with us in the middle. At one point a woman from Ora turned up and began to launch into the unmistakable wail of ritualised mourning. She had been informed by the other village that her sister who lived there had recently died. Her ceaseless guttural howling, which lasted several hours, added even further to the electric tension between the two villages. It felt like an atmosphere that could bubble over into something awful any second, and we tiptoed around organising our equipment, heads down, trying to avoid eye contact with any of the glowering machete-wielding men.

Eventually Jonny Keeling took charge and managed to convince representatives from the two villages to get together and chat.

This too started badly. The senior Ora representative was an unusually tall, skinny, bearded shifty-looking guy, his *bilum* stuffed to bursting with our stolen food, and his teeth stained blood-red with betel nut so he looked as though he'd been feeding on fresh babies. I'd be as likely to trust him as a roadside hitch-hiker carrying a blood-stained chainsaw and a six-foot chunk of rolled-up carpet. In a mild irony, the Big Man from the other village was one of the smallest men I'd ever seen, reaching barely to my stomach. He made up for this lack of stature, though, with a dramatic ferocity, waving around a machete as long as he was and with a fearsome scowl on his face. Whatever he was yelling was clearly intended to convey that he was very unhappy indeed and would not take lightly being cheated out of what was his. Both men looked like big trouble waiting to happen.

There were now perhaps a hundred people gathered, and it had started to take on the feeling of a bizarre refugee camp. There wasn't really room around our tarps and kit tables for that number, so they took to hacking away at the forest to clear space for themselves to sit and also to give them the chance to swing their mighty machetes. We'd gone to great lengths to preserve the biggest and most beautiful trees, to make sure our imprint on the area was as slight as possible. The village folk, though, seemed partly to be chopping them down for something to do. This was all a bit weird, especially as to begin with they were claiming that one of our greatest crimes was chopping down the trees of their sacred grove.

The negotiations were protracted and hot-tempered. Much of what went on we really didn't understand at all, but it was clear from the yells, the pointed fingers and the brandished machetes that neither side was happy. Both Ora village and the newcomers were *wantoks* ('one talks'), which is pidgin for groups of people that share a common tongue. There are a recognised 841 languages in Papua New Guinea alone, while the average number of speakers per language is less than seven thousand. Many of these languages are spoken only by a single isolated village of maybe twenty or thirty people. Can you imagine how isolated your life would be if you could only communicate with the twenty people you lived with? It's no wonder that life here is so splintered, and the people so protective of their boundaries.

There is nowhere else in the world that has such a dizzying diversity of languages, so to be a *wantok* is a piercing, penetrating concept, uniting people that might otherwise be totally alien to each other. The villages had, as they say, 'made ropes' between each other, links of marriage and then birth. Our wailing bereaved woman was a perfect example. A conflict between the two places, then, would be a big deal. A minor skirmish would be an impossibility: it would either be nothing at all or all-out war.

Jared Diamond wrote in the *New Yorker* in 2008 of a clan conflict he witnessed here that had resulted in bloody eye-for-an-eye retribution battles that lasted a generation. He noted: 'Nearly all human societies have given up the pursuit of personal justice in favour of more impersonal systems operated by state governments.' But here in New Guinea, 'without state governments, war between rival groups is chronic'. One of the highlanders he interviewed, who had just organised a successful fatal attack on a rival clan and thus was himself a target for a revenge killing, was unapologetic about the brutal nature of New Guinean justice. He also said: 'I admit that the New Guinea Highland way to solve the problem posed by a killing isn't good ... we are always in effect living on the battlefield.'

Certainly, my earlier experiences in West Papua had borne this out, when I encountered the last member of a clan, the others wiped out by neighbouring tribes. Peter Matthiessen in his seminal book *Under the Mountain Wall* writes at length about the highly ceremonial nature of such battles, and the conventions they follow. Clans gather on purpose-cleared battlefields and start the mêlée off with chants and shouted insults, before throwing missiles from a distance, then finally approach to make the kills and spill the blood necessary to placate the clan.

This is perhaps what makes New Guinean anthropology both so fascinating and so frightening. I've read thousands of pages about this complex and complicated place, and not one of the writers who came here left without experiencing a brush with violence of some kind. In reality, though, this New Guinean way is just how we would have lived our lives before culture and society got in the way of our nature. Perhaps we've not entirely left it behind ourselves. I can vividly recall as a child standing protected by my father at the visitors' end at White Hart Lane when Spurs played Manchester United. The two gangs of supporters, separated by huge wire fences, screamed insults at the enemy and bayed for their blood, before letting loose a volley of coins and other missiles. Afterwards there was vicious fighting outside the ground – tribal violence ritualised within Western civilisation.

Meanwhile, under the tarps, Moose was plotting all of the new kilometres of cave we'd discovered into a laptop, and creating a three-dimensional profile of how the cave might look. The laptop was a pretty space-aged piece of technology even for us, but for the scores of assembled locals it must have seemed sent from space. Banks of faces crowded around the screen, wide-eyed and transfixed. The mature women clearly had more gumption than anyone else, and would barge in to appropriate the best spots, using their elbows to admirable effect in order to get right in by Moose's side.

Then Johnny Young started playing back some of the rushes on the monitor, and it all went crazy. There must have been fifty people crammed in under the tarp – it had been a squash for just the ten of us. They were standing all over the beds and camera cases, jostling for a glance at the weird magic the white men had brought with them.

This perhaps is a taster of how the first 'cargo cults' came into being, when traditional tribal societies first came into contact with technologically advanced explorers in the nineteenth century. Tribes in New Guinea and Melanesia became obsessed with obtaining the 'cargo', the booty they saw being brought into their world. However, they had no understanding of the history that had seen the development of such goods, or of how they could be created. For them, possessions consisted solely of the things they saw around them, and could only be garnered either by collecting them yourself or trading with others. They saw white Big Men arriving in planes and huge boats, who did no work, gathered nothing for themselves, yet possessed luxury cargo which surely hadn't grown on trees or been hooked out of the sea?! They assumed it was the result of great favour granted them by the gods that had allowed these white men their cargo while they, the tribal peoples, had been left with nothing.

Village shamans and Big Men devised elaborate 'myths' that described this cargo as being *their* rightful possessions, somehow misappropriated in the spirit world by the white men but at some time in the future destined to return to them, its rightful owners. The cargo cults thus developed rituals and sacred practices that appealed to their ancestors and to the deities in their forests to bring them the worldly goods that were due to them. The cults flourished particularly around the Second World War, when all of a sudden Japanese, American and Australian soldiers turned up in New Guinea with, it was thought, nefarious unexplained intentions and unimaginable amounts of riches.

When the war finished and the bases closed, the soldiers, navy and airmen departed and all these goods simply disappeared. Totally baffled at this course of events, many tribes started to build offices, airstrips and tanks out of wood. They carved wooden headphones and built radios out of coconut husks in the hope that this would entice the cargo to return. Entire tribes would dress up in warpaint that mimicked the uniforms of the Western soldiers, as well as performing drills and marches with sticks as rifles, so that cargo would fall from the skies as it had when airdropped in by parachute for the fighting forces. They built straw aeroplanes and stood on airfields cut from the forests, waving symbols, believing this would make the planes land again.

In our negotiations, nobody wanted to budge an inch. The second village, as mentioned earlier, wanted all of the money and goods that had been promised to Ora to be given to them, and for Ora to receive nothing at all. It seemed almost as important to them that Ora get no money as it was that they should get paid themselves. And all the while that the two Johnnys, Dave Gill, Wesley and Thomas were entrenched in their task of preventing all-out war, we had to be constantly vigilant. Now, not only the newcomers but also the folk from Ora were openly going through our stuff and taking what they fancied. We pretty much had to sit on our equipment to prevent it being nicked.

We had quite limited bargaining power, just a chunk of money in the local currency *kina* (*kina* is the name for the pearl shells that were used for trading before a more organised currency was introduced).

This amount of *kina*, all of it, had been promised to Ora. Our other expenses were already covered, and in fact we hadn't expected to be spending anything out here in the forest! Other than that, we had ropes, the generator, a bit of leftover food – not much, really. Every half-hour or so, the tiny Big Man would leap to his feet and start yelling, eyes white and wild, swinging his machete.

Keeping cool was next to impossible.

Eventually though, as Tim and I were coiling the last of the ropes into sacks, a huge unbridled cheer of joy went up. Everyone was leaping up and down, waving their machetes in the air, before dissolving into a chorus of 'Whoop whoop whoop!' The little Big Man was pumping Jonny Keeling's hand, and even the lanky Ora man was grinning so wide it looked as if his head might flap right open.

Jonny explained: 'They've agreed to split everything right down the middle, each gets half the money and half the goods we leave behind. We'll also hire our porters to carry the stuff back equally to each village.'

'Jeez, JK, how on earth did you manage that? The guy from Ora's getting half what he was expecting, but he still looks happy!'

'Well, I reckon he knew he was in the wrong. They were just hoping the other village wouldn't find out about us.'

'He got busted, basically?'

'Yeah, and I think he knew that if he didn't agree, it'd be war. He may have got less, but it's still a fortune, and they get to have peace for a few more years at least.'

Peace, it seemed, was cause for great celebration. The mood had changed from charged and dangerous to ebullient – even ecstatic – and all in the space of a minute. It was a noisy, grinning crowd that bore us back to Ora, and the magic whirlybird from the sky that would bear us and our cargo out of the jungle.

The expedition had begun well, though it had taken more out of us than we had expected, and probably more mentally than physically. We were to return home to some decent food and clean clothes – to be dry for a few weeks and prepare for the trip's main expedition, which would take us to the central highlands of Papua. I would be reunited with the main team from *Expedition Borneo* and *Guyana: Lost Land of the Jaguar*. It was our opportunity to discover new species and drum up the support to protect the forests of New Guinea for future generations. Also, our chance to explore a true lost land within a volcano called Bosavi. And none of it, I fervently hoped, would be underground.

11

On Saturday I broke my back.

It was one of those Wimbledon-ruining summer days we love to whinge about in Britain, weather taunting us with bright sunshine then dashing our hopes with sudden showers. The green oaks and yews of the Wye Valley gleamed with diamond drips and resounded to the chink-chink of startled blackbirds. So Tarquin Cooper, my climbing partner, and I had decided to beat the crowd by heading to a quieter crag. He's already had a lifetime of stick for his posh Christian name, so I'll not add any more here. Suffice to say, some of my favourite and most epic mountaineering expeditions have been done in his company. We've learned the sport together and he's a trusted friend. In a year packed to the gunnels with expeditions and trips filming exotic animals, this was the only weekend I had to myself to enjoy warm rock and the familiar natural wonders of a British summer.

We had returned after the New Britain caving escapade to give ourselves a little time to recuperate and prepare for six weeks of torment and torture back in the tropical rainforest. Time here in civilisation was short and precious. Perhaps the desperate desire to experience a whole summer in one weekend was what really caused my lack of judgement. It's possible that my climbing partner and I were over-eager in judging the rocks dry enough to climb, despite the summer showers. Certainly, the rock held my fingertips and my rubber rock shoes, and as advice drifted up from below I lay my weight back against a fissure in the face and levered myself above the trees and into the sunshine. Jackdaws cawed in annoyance at the invader, as I stood spider-like on the vertical rock, body weight balanced tippy-toe on invisible imperfections.

It is in the Wye Valley, in one of the most glorious places for outdoor enthusiasts in Britain, that England and southeast Wales meet. Our original intention had been to climb at Wintour's Leap, my favourite section of rock within easy reach of home. Unfortunately. the weekend sunshine had brought out legions of climbers, including crowds of red-helmeted, irritating and occasionally dangerous novices, so we'd decided to beat the crowd by heading to a quieter crag. The guidebook listed the spot we'd chosen as 'having gained a reputation for loose, dirty climbing ... some virtually unclimbable and dubious rock ... *extreme caution* is advised'. Some sections were closed off as too dangerous to attempt. This did mean, though, that we had the place entirely to ourselves and could take our pick of whatever section of rock we fancied. We'd spent all day finding our way up the choicest routes on the limestone cliffs in blissful solitude, and inevitably as the afternoon drew on and the fritillary and tortoiseshell butterflies lazily fluttered about my ears, I decided to push myself and take on something that would really test me.

When I tell people I'm into mountaineering, their usual response is 'Ah, one of those adrenalin junkies, are you?' But I really don't see it that way at all. Climbing is not, fundamentally, about risk-taking, about dicing with death and being careless with the priceless commodity that is life. Truth is, anyone who was into climbing and was truly reckless would not last long. In fact, I think climbing, and indeed most so-called extreme sports, are about exactly the opposite facet of human nature.

It's about being in control. It's about knowing that your own abilities and experience will allow you to do things that mere mortals couldn't contemplate but that for you are achievable entirely safely. In essence, it is about being able to take your destiny in your own hands, about being able to put aside the constrictions and regulations of the modern world. Living in the cosseted twenty-first century we are constantly being told what to do, always protected from perceived dangers and having responsibility for our own survival taken from us. Many people find that as big companies, government and the media seem to virtually take over society, we can have independence in precious few areas of life. We seem to be pawns in a massive game with no say over eventual outcomes.

This just isn't true when you're climbing. You can simply take a small rucksack and stroll off into the hills, with total freedom over your actions, taking yourself off into a world that's yours and yours alone, even in overpopulated Britain. In this, it is the most liberating yet disciplined of arts.

The best rock climber I have ever had the privilege of partnering is a superhuman Brit called John Arran. John is a freakish athlete who once set the standard for the sport here by setting the hardest rock route in the United Kingdom. He made the first ascent of the Angel Falls in Venezuela and has achieved numerous other internationally significant firsts around the world. There are many in the sport who plaster themselves all over the media and have egos the size of the peaks they scale, but no one would ever know, from John himself, how special he is. He's extremely humble, would never crow about his achievements, and is always ready with a – completely unpatronising – helping hand, even to a Sunday-league hacker like me.

You could never ascertain how good he is just from watching him climb. He has such indescribable precision that he walks up an overhanging wall with no visible holds *as if he is climbing a ladder.* No frills, no flouncy moves, no grunts, groans or flexing, in fact no show of effort whatsoever. It is the most extraordinary display of control you will ever see from an athlete.

A few years back, John climbed a hundred routes in twenty-four hours, every one of which was at a level that, even equipped with all the ropes and modern kit imaginable, I would struggle to ascend. John climbed them all solo. Without a rope. This seems like taking the most ludicrous risks, until you become aware that John is so incredibly capable that those climbs are to him as easy as a mere scramble would be to you or me. He has enough strength and technique in reserve that he can test every handhold to ensure it won't come out, and the judgement to gauge when to downclimb if it looks sketchy. He is not going to fall, is operating utterly within himself, and thus in total safety. Far from being a dice with death, this is the ultimate freedom, the ultimate mastery of one's own fate. While I could do nothing but rock climbing every hour of the rest of my life and not come close to the likes of John, still I strive for the same purity of purpose.

There is nothing as liberating or as empowering as moving comfortably up vertiginous vertical faces. The mere act of leaving the earth and taking hold of warm rock or blue ice sets you apart from the rest of the ground-bound populace, and sets you free. On the other hand, there is nothing so crippling, so utterly terrifying, as finding yourself out of your depth, exhausted, and in mortal danger – it is paralysing. I don't deliberately put myself into situations like that, but when you continually challenge yourself without necessarily having the levels of control you aspire to, things occasionally go wrong.

We'd just had a quick ten-minute burst of rain that'd sent Tarquin and me scuttling for shelter under the conifers that grew right up to the edge of the rock. As soon as the rain abated, the sun came out and the ancient limestone started to steam and dry out. It was my turn to lead, and I was keen to hit something hard that would feel like my big achievement of the weekend. Having cruised through the last few climbs I was feeling strong and confident. It was also an oddly aesthetically satisfying piece of rock, with a perfect longitudinal crack above a jagged ledge, breaking out above the trees and into the sunshine.

'Are you sure about this, Steve? I think this one looks really sketchy – horrid, in fact.' Tarx was far from sure about it.

'Yes, I think so, it looks like there's some bridging up there. Just my type of climbing, and it's only one grade harder than that last one – we sailed that!'

'Right . . . but the guidebook says it's "very hard for the grade" – says here it "pulls no punches"! You sure about this, mate?'

'Damn right, Tarx! Me strong like bull! I'll knock this one off, then we can go find somewhere to bivvi – reckon we'll have earned some fish and chips!'

'Bloody hell, Backshall, just don't blame me if I have an epic following you!'

This was characteristically modest from Tarx because he is at least as strong a climber as me. In fact, being at a similar level of accomplishment and ambition is perhaps the most important facet of a good climbing partnership. That and being able to sit together in a tent in a blizzard for days on end without actually throttling each other.

There were no nerves to begin with, just the certainty that this would be a very satisfying accomplishment to go home with. However, I had misjudged it quite badly. The crack itself was still damp inside, with slimy green moss and goo providing precious little traction. A big chockstone boulder, a crux hold for the climb, wobbled alarmingly and had to be avoided. Mere metres from the start, and my fingers and calf muscles began to quiver with the effort. I started to puff and blow, as much a sign of growing alarm as of exertion. However, the beauty of climbing grading systems is that usually if you're climbing something that's placed in a grade you are capable of, eventually it will come good. A big hold will appear out of nowhere, or the gradient will turn out to be an optical illusion and it will suddenly become a piece of cake. I pushed on, in the certainty that this was going to happen – any second, that magic handhold would appear. Any second now.

It didn't appear. I was perhaps thirty feet from the rocks beneath, my forearms fizzing, hands and fingers starting to tire. There was a choice to make. I could push on to a small ledge perhaps a metre above me where there might be a decent hold. However, that would take me above my camming protection. If I fell, the forces loading it would be dramatic and it might not hold. My snap decision was slightly coloured by fear and exhaustion: actually to carry on to the hold would have been by far the safest thing to do. Instead I decided to do something many climbers do as a matter of course, but I never have: I lowered myself slightly and sat back to rest on my rope, which was secured by the camming protection I'd placed into the crack before me.

So-called traditional climbing relies entirely on cracks in the rock. Back in the early days, tough Yorkshire welders and Welsh mountain men jammed machine rivets, knots of hemp rope and iron bolts into the cracks, then when they were good and wedged tied their ropes around them in order to protect themselves. Nowadays we have expensive camming devices called 'friends', made of the most lightweight modern materials, to do this job. They open up inside a crack, and take your force, should you fall. Unfortunately, even the new modern gear has to be placed correctly in order to work, and even a perfect placement can on occasion fail.

Tarquin's voice came up to me with a note of urgency I'd not heard from him before: 'Are you *really* sure of that friend, mate?' I remember a feeling of mild annoyance – of course I was sure of it! Was he suggesting my gear placement wasn't up to scratch?! Turning to look down and give him a jocular piece of my mind, I stopped dead in horror. Every piece of protection I'd placed below me had lifted out of the greasy crack, and was now swinging merrily on my rope like little Chinese lanterns, as useless as a chocolate jockstrap. My entire bodyweight was being held on the single friend in front of me, no thicker than my little finger. And it was a very, very long way down. Ever so gently, I reached up to get myself back on to the rock, willing with every cell in my body the friend that was holding me to stay in place. It rocked towards me with a nails-down-blackboard squeak. I will see in nightmares for the rest of my life the second it popped like a champagne cork out of the rock, my rope, my lifeline, going slack as I pitched back into space.

People talk nonsense about their lives flashing before their eyes at such moments. It did, though, take an incredibly long time for me to hit the ground, a blur of flashing rock and trees, enough time to realise what was happening and that it was going to mean big trouble – and then an instant heavy deceleration. It felt like being hit by a truck. Three storeys' worth of drop, windmilling my arms to keep myself upright, and I hit the jagged rock ledge with one foot, the force of my entire bodyweight powered down through that single leg, spinning me backwards to drop another three metres or so to the ground.

There was no righting myself this time, and I landed flat on my back, making an almighty concertina of my upper body, driving the air out of me like a massive punch to the kidneys. There was no welcoming unconsciousness, no deadened sensation from the force of the shock, just the most extraordinary pain I've ever experienced. It was instantly clear this was going to be a very bad day indeed.

Tarx had a hold of my hand – 'What've you done, what've you done?' – and was pulling my climbing shoe off to reveal a grotesque gaping flap in the underside of a barely recognisable foot, while I screamed from the gut, a mixture of anger, shame, and that unimaginable pain. The force of the fall had driven my heel bone down clear through my foot, and it looked as if someone had hit the sole with a hatchet, the cleaved halves of flesh revealing the bone beneath. The incredible force had dislocated my ankle and broken both my ankle and my leg, and the chasm in the underside of my foot was flooding blood. Tarx was incredible, crystal-calm, unclouded by the taint of emotion. He urged me to sit still, and rushed to his backpack for water and the first-aid kit. He'd brought a first-aid kit for a weekend's cragging! How often am I organised enough to bother with that?! He couldn't hide the disgust on his face, though, as he flushed out the wound in my foot with water. He then packed it with padding and bound it as tightly as I could bear, and got out his mobile to call for the cavalry.

'Right mate, we've got a problem.'

'No shit we've got a fucking problem! My heelbone's sticking out of the bottom of my foot! I'm supposed to be going away filming in three days!'

'First things first, mate. There's no mobile signal up here.' He paused a second to let this sink in. 'I mean, I can go down to the road and try and flag someone down, I suppose, but let's face it, they won't be getting a rescue chopper in here with all this forest anyhow. We're on our own, buddy.'

'Tarx, it took twenty minutes to walk up here – I'll never get down there like this.'

'Steve, we've got no choice.'

Then something occurred to him. 'Your back's OK, right? You came down on it like a fucking sack of coal.'

I pondered briefly. It felt as if I'd been given a good kicking by the entire England rugby team, and every bit of me was in agony, but the foot was the thing I was focused on – it was what medics call a 'distraction injury', big, bloody, gruesome, distracting you from more subtle but potentially more dangerous problems.

'The back's fine, Tarx, let's just get me down this damn thing.'

Tarquin took both our backpacks and ran off down the hill a few hundred metres, dumped them, and then came back up to sort me out. He found a big stick I could use to prop me up on my good side, then slung my arm over his shoulder. Tarx is a seriously tough cookie, but much lighter than I am, and couldn't keep me upright the entire way. Every time my cloven left heel touched the ground the pain took the strength from me and I dropped to the ground, sitting flat on my backside retching with agony. It took well over an hour of this unimaginable torture to get me down to the road. An hour of agony that makes me feel sick even now if I think about it.

From there Tarx ran to get the car, and drove me back over the border to England, and hospital in Bristol.

Obviously, if we'd realised I'd broken my back in two places, we'd probably have handled it differently, but we were so fixated on my foot that the back seemed fine by comparison.

But it's not fine. It's not fine at all. And so I find myself in the unusual position of having a long time ahead of me in which I can do nothing, and I'm not good at that. Nothing to do but write and think about how I got to be in this predicament and this place. The pain and the strangled sounds of the old man in the hospital bed next to me trying to die are too much to bear, so I'll just press the button that gives me intravenous morphine on demand, feel the pain blissfully drift away, and sleep.

12

Rain streamed up the glass in front of my face, dense cloud the only thing visible before us. It felt like being in a submarine, surging through a milky sea. Beside me, the mirrored sunglasses of my pilot glinted white. He seemed remarkably calm, so I supposed there was no danger of us flying face-first into an unseen cliff. Then with a sudden change in pitch from the rotors above, the helicopter dropped from the rain clouds into a forested gorge cut through by a fierce caffè latte river. I felt the familiar excitement return. The forest beneath me laces the foothills of a remarkable range of mountains – limestone peppered with million-year-old volcanoes. The terrain is sliced by mighty rivers, the rainfall of millennia having helped to create untold cave systems and more waterfalls than I've ever seen. The karst limestone is decked in the most dramatic flora of any tropical rainforest. Immense ferns, spindly palms, tangled strangler figs dripping with orchids and other epiphytes. Like getting a window back on the landscapes the dinosaurs roamed.

It was seven months after the fall that nearly took my life, or at least my ability to walk, and finally I'd shaken off infirmity enough to return to the forest. Given that my recovery was so protracted, I was gobsmacked that they were prepared to wait. If it'd been me in charge of the programme, I would have ordered the expedition to continue and found a new presenter. Luckily my colleagues at the BBC were not so mercenary.

While I was still in hospital on a drip, catheterised, drugged to the eyeballs and in a full body brace, series producer Steve Greenwood and executive producer Tim Martin came in and promised they'd wait for me to recover, whatever it took. At a very low time in my life, this was one of the shining lights. I'd come to lose faith in the shallow, back-stabbing world of telly, where loyalty was rare indeed. (That said, not wanting to push the boundaries too far, I've never mentioned to Steve that when he came in on a later occasion to discuss things, under the sheets I was sat on a bedpan trying desperately, and failing, to have my first poo in over a week. Nobody ever tells you what opiate painkillers do to your bowels, that they back you up like you've got a cork up your bottom. Steve stayed and chatted for nearly an hour, completely oblivious (I think) to the cause of the odd squint I must have been sporting.)

My broken back mended remarkably quickly, and it seemed I was going to be up and running in miracle-quick time. My ankle, though, had other ideas. Not even the surgeons had really appreciated how much cartilage had been destroyed, effectively taking all the shock absorption and lubrication out of the joint, and the bones no longer aligned neatly as they should. The ankle, it seems, is a more complex and less fixable joint than just about any other. I had five operations to try and set it right, including in later months having to wear a metal cage, an instrument of torture, bolted through my bones, before it became abundantly clear that it would never heal. Ahead of me was a lifetime of pain. Running and jumping would probably be out of the question, and even walking would never again be pleasurable. However, there was no point getting all whingey about it, and I had some serious issues with New Guinea that still needed to be addressed. I gave Steve and Tim the OK, packed my painkillers, then boarded a series of planes, helicopters and boats for the long journey back out to the jungle. I still had to get the jungle out of my system.

There were to be satellite trips all over New Guinea, but our basecamp and therefore the place we would be spending the most time was in the Southern Highlands. The lowlands in the south rise sharply through limestone peaks, gullies and gorges to volcanic peaks in the centre and north of the province, where Papua New Guinea's second-highest mountain is located. Outsiders didn't reach the southern highlands until 1935, and even today it's not exactly Benidorm, with few foreign visitors venturing into its wild places. The overwhelming majority of outsiders to be found there are connected with the Australian energy giant Oil Search (whose majority stakeholder is Exxon Mobil). The area is particularly rich in resources, with potentially some of the biggest oilfields in New Guinea, and has become a focus for this company, not always with the most harmonious outcomes.

Despite the fact that oilfields here are expected to raise as much as $11 billion, there is not yet any palpable improvement in the lot of the people in the region. And there has been increasing violence, discontent and political instability amid rumours that the relationship between the government and foreign multinationals set to exploit the land is too cosy. In fact, just three years previously, the Papuan government had declared a state of emergency, claiming that lawlessness, instability, theft and corruption were jeopardising further operations. Illegal guns are plentiful, and the murder rate is equivalent to that of Colombia and some of the worst US cities. Soldiers and police swamped the area, declaring curfews and aggressively bringing down any resistance. Even today, travel advisories to the southern highlands set it apart as one of the most volatile places in Papua New Guinea: it is 'subject to inter-tribal disturbances without warning ... you should exercise extreme caution, as law enforcement in these areas is weak.'

The helicopter soared down the length of the river gorge, driving back the trees with the blast from the rotors, its skids seeming to skim the surface of the brown and occasionally white flow beneath us. After a spectacular journey that could have gone on for ever as far as I was concerned, we rounded a corner and hovered above a rocky beach jutting out from the steeply forested banks. In amongst the trees at the riverbank I caught my first glimpse of the group of green tarpaulins that would be our home. Their framework had been sourced from local materials, with all of the mighty tropical hardwood trees left intact, so even when we were in camp we were engulfed in the claustrophobic embrace of the forest; touched by its tangled tendrils, deafened by the voices of its bugs, frogs and birds.

As we landed, entomologist Dr George McGavin crouched beneath the whirring blades to come and welcome me, hand outstretched, a grin on his face. I noted that he had lost weight, and was looking rather dashing and fit; and registering his coordinated khaki jungle gear and perfectly trimmed facial hair, I felt like I was being welcomed by a film star! Behind him on the beach, the whole team stood expectantly. Some were fond friends and respected colleagues waiting with hugs and good wishes, others were new faces brought on to the

New Guinea team for their specialisations. It was so overwhelming that, without thinking, I threw my backpack on to my shoulder, forgetting the snake sticks I had tied to the side of it and the too-close rotor blades above ... a close call – that would have been a very embarrassing start to the trip! Within ten frantic minutes I had renewed old acquaintances and met the newcomers who would be my constant companions for the next five weeks.

The forest here was as dramatic in its flora as you will see anywhere in the world. Everything about the place seems to have been plucked from prehistory – it's like a dripping, steaming Carboniferous kaleidoscope, supersized and sublime. There were times when you'd not have batted an eyelid had a pterosaur swooped out of the verdant velvet cloak about us, or a velociraptor growled from the undergrowth. There was no need for hyperbole in our descriptions of the place. It was unexplored and unplundered by outsiders, genuinely remote from anywhere, a daunting primeval place ripe for exploration. Our main aim was to document the plants and animals that live there, and the paucity of biological expeditions to the area in the past would mean our labours would certainly result in species new to science. The main goal of our scientific survey was to try and convince the New Guinean government to put more effort into protecting this area of such exceptional diversity and of untold significance but in severe danger of over-exploitation.

The basecamp was situated on the lowest slopes of a million-year-old volcano that dominates these highlands. Covered in jungle that has never been penetrated by any outsider, Mount Bosavi is seven times the size of Vesuvius. Even the few clans that live on the mountain's outer slopes only venture inside every few years, when the immense effort of entering the cauldron is rewarded by the superb hunting offered by its unplundered heart. It was to be many weeks, though, before I even glimpsed the wondrous peak, as it was many miles from our location, its summit almost permanently shrouded in cloud.

As with all the preceding expeditions, our main basecamps are rather bizarre places. Television cameras are temperamental beings, as prone to fits of pique and tantrums as spoilt children, and the things they hate most are humidity and rain – which are the defining characteristics of the rainforest. In order to keep the beast from the machine and actually be able to make the films, basecamp has to be a well organised and civilised place, with electricity for at least parts of the day and dry places where the kit can be pampered. Consequently, basecamps always look a bit too – well, nice, a bit like eco-lodges, and every middle-class kid who's been on a gap-year sojourn on Operation Raleigh will be sat watching the fruits of our trip at home indignantly exclaiming, 'Bloody luxury!'

Being as we're supposed to be making a film about suffering and struggle in the jungle, everything shot in basecamp has an air of artifice about it. Someone stubs their toe and two seconds later a film unit is mobilised, there will be endless shots of medics shaking their heads, and you can hear in your head the voice-of-God narrator talking in a sombre tone about infections, wounds crawling with maggots and evacuations.

In fact, the closest I come to death in the first weeks is when trying to swim across the fast-flowing river in my new jungle trousers. The huge hip pockets (which I thought looked cool) fill up with water like parachutes fill with air, and start to carry me downstream. This would be an embarrassing way to go. Striking out for the bank fiercely and somewhat desperately, I briefly consider popping the fly button to release me and just letting my trousers disappear off downstream. However, like most of the guys on the crew, I spend my jungle days going commando – that is, free from the constriction of undergarments. The cameras are rolling the whole time, and I would have to walk back into the camp as naked as the day I was born. That would live in the annals of expedition history for all time. Weirdly, this idea panics me rather more than the thought of drowning, and I strike out desperately, forgetting my nice freestyle form and going into swimming survival mode. Finally my feet touch the sandy bottom, and I drag myself ashore.

After being woken in my hammock three mornings running by a burning light and a camera in my face, I'm desperate to get out of basecamp. All this is, of course, in stark contrast to the so-called satellite trips I've been privileged to embark on with my small team of toughies: first ascents of virgin summits, rope forays into the roaring heart of frozen glaciers, descents to the bottom of uncharted waterfalls, abseils into unexplored sinkholes and the cave systems beneath – all of a sudden, things can be a bit *too* real. Once you start to look at the realities of sending people who may have families and kids into remote and very dangerous environments where there is no evacuation plan and literally anything could happen – well, let's just say I'm glad I don't have big boss Steve Greenwood's responsibilities. It's no wonder he spends half his day standing alone, deep in thought, face slightly contorted with concentration, his hands pressed into the small of his back. This is the kind of pose Custer might have adopted, just after being told there were rather more Indians out there than expected.

The best function of basecamp is traditional natural history filming in an area that no one really knows anything about. On a conventional natural history film, if cameraman Gordon Buchanan was wanting to get a sequence with birds of paradise, he'd go to the best place in New Guinea to see them, and set up his hide at a *lek* (their dancing site) that's known to be in use. Even with this degree of forethought and preparation, he'd probably still have to wait a month to film a sequence. Here it's even harder, as we're starting from zero knowledge and have to make it up as we go along.

Gordon's first challenge is to film cassowaries, which turns out to be way harder than any of us had imagined. We very quickly come to the realisation that although these forests have never been explored by white people, they have been very well hunted over generations by the locals. Animals are thin on the ground, and those that *are* around are very skittish and scared of people. The best Gordon manages are a few grainy night shots on a remote infrared camera, but not enough for anyone to be satisfied with. His next task is to film megapode birds. These are relatively common throughout Australasia. They make a huge nest of leaves, which as it decomposes creates warmth that incubates their eggs. I'd filmed them before in Australia, lying right on top of a nest and with the megapodes pecking up food inches from my face. The footage never made it to the screen, as frankly it looked like a couple of half-plucked turkeys on a compost heap. Imagine, then, Gordon's delight, not only at having to spend a week trying to film these birds, but even worse, them not turning up and him having to give up on the whole thing.

Finally, he has to turn his attention to hornbills, which fly noisily overhead all the time but are harder to film than dodos. Every day he is up and out at 4 a.m., and back late at night wet and muddy, having sat in a hide covered in sweat bees all day. Three weeks in, and even the chipper Scotsman is starting to look a little crazed.

Finding these lowland forests to be quite low on wildlife was a bit of an unpleasant surprise. It proved that local people are capable of hunting out areas that are quite a distance from their villages. This doesn't bode well for the future of New Guinea's wildlife, as the human population increases and the forest cover decreases. These creatures that rely on the forests will be subject to what biologists call 'feedback loops' and 'tipping points'. Imagine a graph showing deforestation, human population and the hunting of wild animals, all of which are increasing at about the same rate: three neat coloured lines going up at the same pace. Predictions based on this graph will show that if human populations keep growing at the same pace, and forests are being cut down at the same rate, then we'll lose the animals at the same speed too.

However, what actually happens is that when degradation reaches certain levels, things accelerate, go berserk and start to fall apart far faster than expected. In this case, the forests become fragmented into little pockets, the animals get separated from other populations and have nowhere they can go to escape hunting. The people, now with guns as well as dogs, can cover every inch of forest, their feral cats, dogs, pigs and rats run rampage in the woods, and in the space of a generation *everything* disappears before anyone has a chance to realise it's happening. In his wonderful book *Throwim Way Leg* (the pidgin phrase that describes the start of a journey), Tim Flannery comments with his customary wisdom on how local people that we might idealise as hunter-gatherers working in perfect tandem with their environment can often be anything but:

> The development of conservation programmes in countries such as Papua New Guinea is fraught with difficulty. Western notions of conservation often appear to be completely nonsensical to the local people. Many people believe that the animals of the forest have always been there and will always remain. When faced with clear evidence of a decline in abundance, or even

> extinction, they will point to a place over the mountains saying, 'There's still plenty over there.' Little do they realise there is always a village 'over there', inhabited by people who when asked the same question, point back in their own direction.

Basecamp provides not only the shots of the animals, but also the real scientific discoveries. A segment of camp has been set aside as a jungle lab, thick with the heavy but not unpleasant smell of preserving alcohol, and is inhabited by peculiar and wonderful beings that never fail to fascinate. And I'm not just talking about the animals. Clutching bagfuls of living wonders, the scientists scurry about their little domain, mumbling about sexual dimorphism and predicted evolutionary lineages and other rarefied things – these are human beings with brains the size of Wal-Marts and possessing all the associated quirks and queernesses that seem to occur when people fill their heads with information and leave no room for the normal stuff.

Allen Allison is a herpetologist who knows the scientific binomial name of every creature since diplodocus, and can quote you the author and date of every paper written over the last hundred years (honestly – it's utterly spooky). Yet he regularly walks into the lab three or four times over and looks around him, blinking blankly, in a way that tells you he's thinking: 'I'll be darned, what did I come in here for?' Cameraman Gordon Buchanan suggests that like many people who spend too much time with one kind of animal, Allen has become a bit like his subjects, though this is based solely on the fact that he speaks with a high-pitched American accent and sounds a bit like Kermit the frog.

We all like him enormously. His work is (particularly to me, fascinated as I am by anything to do with herpetology) extraordinary. He tracks frogs using the male's evening advertisement calls, in order to record them up close with high-quality microphones. As most species will stop calling as soon as they think a potential threat is nearby, Allen may spend several hours just creeping close enough to one animal to get a clean recording. Using special wave-form software, he can then predict how closely related animals should be differentiated into new species, based solely on the differences in their calls. DNA evidence has proved him right time and again. He's a proper scientist.

Just an hour or so from camp is a shallow black pool that Allen has particularly favoured in his work. It's no more than a squash-court-sized pond, but is absolute frog heaven. This may seem weird in a rainforest where water seems to be everywhere, but there's good reason. The main river flows far too ferociously for tadpoles to stand a chance. The ground is so porous that puddles last no more than a few hours, and there seem to be far fewer bromeliad pools in the trees than you'd expect in a comparable South American forest. Because of this, small semi-permanent pools like Allen's are prime real estate for frogs.

The first time I head to the black pool after dark, I'm struck dumb by the diversity of amphibians there. The water is writhing with tadpoles of all sizes, and there are perhaps fifteen species of adult frog in the trees around it. One species in particular catches my eye. These are the starkly beautiful tree frogs, which throng the air with their calls. Both males and females are abundant around the pool.

One of this frog's remarkable traits is that when the male has met the woman of his dreams and they've mated, she will teeter out on to a branch above a body of water like this, then deposit her egg mass on to leaves overhanging the water. The tadpoles will start to develop in their sticky eggs, clear of the water and of all the predators within that'd love to gobble up the protein of such an egg mass. Eventually, when the tadpoles are quite well developed, they'll break free of their eggs and high-dive into the water below, free-swimming and well able to escape at least some of their predators (which may include other tadpoles). One predator they cannot escape, though, are flies in the *Drosophila* group. The adult female lays her eggs near a tadpole egg cluster, and as the larvae hatch out they feed on the developing embryos. On the first evening I explored the black pool, there were only the singing adult frogs – no evidence of their sticky union.

Three nights later, I was passing the pool again, on a hunt, and had to rub my eyes and do a double-check. Was I seeing things? It looked as though a giant had sneezed right across the pool. Every leaf, twig and branch overhanging it, up to perhaps double my height above the water, was thick with blobs of transparent jelly. There must have been a dozen bucket-loads of this slime, all created in a single night by a few hundred amorous frogs, and all to be gone again in just a couple of days. It was an absolute wonder. There in the beams of our torches, bizarre Christmas trees bedecked with ... snot. If you focused on one particular blob, the larvae inside would react against the light and start to wriggle, tiny black embryo frogs, protected by their precarious position where almost no predator could reach them. According to Kwiwan, one of the local naturalists, the only animal that *can* reach them – and apparently it delights in eating the yummy bogeys – is the cassowary. This may seem a bit far-fetched – cassowaries are thought of as fruit eaters – but actually they'll eat almost anything. In fact, scientists wanting to study these insane dinosaur-like birds will bait their traps with dead rats: it seems they like nothing better than a bit of decaying meat.

If it seems that I'm pushing the point with the dinosaur–cassowary connection, then it's worth noting that modern biologists, reassessing the reptile family tree (actually, it's no longer fashionable to treat such relationships in the form of a tree, but the image will do for now), place the crocodilians, birds and dinosaurs into one group known as archosaurs. These are more closely related to each other, and to a more recent common ancestor, than the archosaurs are to the other reptiles – lizards, snakes or turtles. And while a bird of paradise may not look much like an alligator or caiman, the cassowary is about as prehistoric a bird as you'll ever see. If you saw a knee-down photo of a cassowary's featherless foot, with its heavy dark scales and curved, black raptorial claws, you'd be forgiven for thinking it was the foot of a T-rex. Perhaps even more similar is that of the moa, elephant bird and modern ostrich.

While our expedition's main drive is to find and describe new species of plants and animals, there is an overwhelming sense that the more information we can glean about this place, the better we can understand the problems that face it, the more weight we can give to our final reports, and the more international pressure we can bring to bear on the Papuan government and the global community.

Our secret weapon in this is Phil Shearman, an ecologist who calculated in about two days how much carbon biomass there was sequestered in our entire forest, and produced 3D maps of the surrounding mountains. Standing about six foot two, bespectacled, and clad when out in the forest either in jeans and trainers or in a full-on boiler suit and wellies (if he's doing dirty work), he is without doubt more at home in the jungle than any of the other scientists on this expedition.

He thinks nothing of heading off for a week or so's collecting trip, taking just a packet of crackers and a sleeping bag, quietly going about his business without ever craving the attentions of the TV cameras or, indeed, any recognition for his work. Not bad, really, considering that he collected during our trip over two hundred species of orchid, of which twenty to thirty are unknown to science.

My explanation for his modesty is that Phil is an Ozzie, and has very obviously spent his life being harried by 'tall-poppy syndrome'. In contrast to America where arrogance amongst the gifted is very much expected, in Australia, big-headedness is often considered such a character flaw that if you show yourself to be exceptionally brainy you can expect to be ruthlessly 'ragged', 'sledged' or 'cut down like a tall poppy'. Because of this, Phil is almost painfully humble. He will casually tell you some jaw-dropping fact that changes your life, then laughingly add: 'But of course I'm a bit of a retard, so that's probably all bullshit.' I am in awe of him, and hang on his every word.

Phil's reports told a very frightening tale about the southern highlands of New Guinea. They have received very few visitors over the years, and until recently have mostly resisted the efforts of the loggers and oil companies to ravage them, purely because the terrain is so unforgiving. The unpleasant fact, though, is that new, modern methods of plundering the wealth of the land mean nowhere is safe any more. On satellite maps of the green wonderland around basecamp, logging roads could be seen snaking their way into the deepest corners, and wherever they penetrated, deforestation spread like an infection.

In the 1950s, before the West started to plunder Papua New Guinea, it was the most forested place on the planet. As late as 1972, 82 per cent of the landmass was still covered with pristine forest, the south of the country possessing the most significant and sizeable coastal mangrove forests, climbing over a thousand metres to the spectacular montane forests along the island's spine. Eighty-two per cent is a truly mind-boggling figure. Here in England's green and pleasant land, less than 7 per cent is still in anything even near its original state – and it's people who have changed all that. New Guinea's forests, as I said earlier, are some of the richest when it comes to biodiversity. It's estimated there are 191 species of mammals living there, though new species are being discovered all the time, and 80 per cent of those are endemic, occurring nowhere else.

Many of these mammals are tied to a particular area and habitat, and if the forests go, those species will disappear for ever. For at least the last ten thousand years, the small agrarian population of New Guinea would have hunted the forest depths, but mostly survived with agriculture – cutting small gardens in the forests, using them for a few crop cycles, then letting them go fallow so that the jungle could reclaim them and the soils recover. However, in the last twenty years, the human population has doubled, and the pressure from within has increased massively. At the same time, huge Malaysian industrial logging companies have moved in and stripped the forests bare, most of their exports going to China. These Malaysian companies are richer than God and fabulously corrupt, lining the pockets of politicians while returning a minuscule percentage of their profits in taxes to the New Guinean government.

Of course, the New Guinean government is corrupt too, so almost none of these riches reach the people who actually own and depend on the forests. When you look down from a helicopter into the forests below, you see logging roads switchbacking to and fro through the trees, even where the topography doesn't require it. This is down to regulations stating that the companies can take trees growing five hundred metres, and sometimes a kilometre, from these roads. By simply quadrupling the length of the roads by doubling them back on themselves, they quadruple the amount of trees they can take.

The statistics that Phil managed to garner on his surveys are worth recording here, as they are truly arresting. He worked out that deforestation and burning are releasing 22 million tonnes of carbon every year – that's equivalent to the total emissions of all the cars in Australia. Even more frightening, New Guinea is losing 2.6 per cent of its primary forests every year: just about the fastest rate of deforestation found anywhere in the world. As a comparison, the Amazon is losing a still horrifying 0.92 per cent annually. Phil's best estimates suggest that in a decade, by 2021, 83 per cent of New Guinea's forests will be gone for ever. A decade?! That's nothing! It's the blink of an eye, less time than it's taken me to put this book together! It's difficult to underestimate what would happen to New Guinea if this came to pass. Those forests bind and replenish the soils. Without them, much of the topsoil will just wash away, there'll be huge landslides, watercourses will flood with sediment, and vast areas of land will turn to desert.

While the real scientists go about completing their missions of true global import, my role in basecamp is merely to be the expedition's snake expert. I'm not a member of the science team, as there's little or no chance of me turning up a new species of snake here, partly because snake diversity isn't massive, but also because that would need me to ditch all my other tasks around base and devote myself one hundred per cent to the job. Snakes are certainly the beasts, though, that I have the most comprehensive knowledge of; it's an area of zoology that has captivated me for most of my life.

I can trace my obsession with reptiles back to a trip to Zimbabwe at about eight years old. This was in the days before Robert Mugabe created such chaos in that gorgeous but beleaguered country and all their wildlife became nothing more than potential bushmeat for the starving populace. Back then, Zimbabwe had some of the finest national parks in the world, genuinely wild, and overrun with game. We were there for only about a month, but it had a phenomenal impact on me. I remember clearly our safari guide Siggy. He was probably in his early thirties, wore a battered bush hat over his long blond hair and a rifle slung over his shoulder, and was squeezed into the ludicrously tiny, tight khaki shorts that plague his profession. He seemed to me to be totally omnipotent.

On walking safaris he'd drop down on one knee, crumble some dung between his fingers and surmise: 'Kudu, a male, came this way two hours ago.' Ten minutes later we'd round a corner and the magnificent antelope would be standing right there, regarding us regally. Apart from perhaps my dad, I wanted to be Siggy more than any other of my heroes, and thanks to Siggy I spent my young years fantasising about being a safari guide. For his part, he obviously spotted me as a young mind desperate for input, and confided in me as if I was the only person there. He let me ride in the front of the jeep with him, seemed to be pointing things out to me and me alone, and would make a point of stopping just to show me dung beetles rolling their cargo across the road, or take a huge praying mantis and place her in my hands.

One evening we were returning to camp and had stopped to pick up the kitchen porter who, as there was no room inside the jeep, took a perch on the bonnet. We were going pretty quick down a narrow, sandy road bordered with viciously barbed acacia scrub – almost like a rally car, with the poor guy on the bonnet clinging on for dear life. Suddenly, Siggy slammed on the brakes and the jeep slithered and crunched to a halt in front of a large branch stretching across the road. The poor perching passenger was thrown clear off and into the road. It was only then that the branch curled slightly, then slithered neatly and quickly off into the bushes. It was an African rock python. Even allowing for the distorting mirror of time and memory, it must have been five metres long and as thick as my young waist (not unknown for a female rock python – they'll eat antelope as big as impala!). It was the most horrifying, frightening, primeval, fascinating thing I'd seen in my whole life.

My determination to spend every second of every day out with Siggy had unfortunate consequences, though. One day I stayed out from sunrise to sunset without shirt or sunblock, and suffered the result that with hindsight was pretty inevitable.

I was rushed to the bush hospital with second-degree burns over my shoulders, and they had to cut my shirt off my suppurating back. Even today you can see the scarring amongst the flocks of sun-damage freckles all over my back and shoulders. This effectively put an end to the safari part of the trip, and we returned to Harare to try and get a flight home. We spent a week trying to get on to the evening flight, and in the daytimes Mum and Dad took us to the local reptile park, me covered in calamine lotion and long-sleeved silk shirt, as that was all I could bear to wear. They only intended to take us once, but after that I pestered them to take us daily. We got to handle baby crocs and pythons, and saw black mambas and swaying cobras in big glass enclosures. I was utterly transfixed.

Over the years I have kept many reptile pets, including geckos that lived free behind my pictures and radiators, calling to each other in the evenings across the lounge. I've had plenty of snakes, including a two-metre Taiwanese beauty snake that used to make Great Escapes from its vivarium and go missing for weeks at a time. We had to tell my housemate's girlfriend, who loathed snakes, that he was at the vet's. Eventually he turned up in a shoebox in my laundry room! However, it's searching for cold-blooded creatures in the wild – 'herping', to the reptile geek – that really gets me excited. Whether as a child catching lizards around rock walls on European summer holidays, or as an adult going out with my snake hook on night wanders in search of venomous snakes or tiny chirruping frogs, there's nothing that makes me happier.

The herping around basecamp was simply superb, and offered far greater rewards than the birding or mammal-hunting. The most common snake, which turned up on every single night walk, was a long thin ground snake called *Stegonotus*. Snappy and rapid, it'd bite viciously if it got the chance, before making a frantic break for safety. They were brown and plain, but always cheered me up to no end. Anything but plain, though, was an amethystine python found very close to camp. They are one of the most dramatic snakes found in this region. It gets to be as much as six metres long, but is thin and spindly for its length, and it eats small mammals. The python is named after the unique sheen it has when you turn it against light, like the rainbowing of oil floating on water or handfuls of amethyst clusters. Underneath that lustre is a divine leopard print in dusty and deep scarlet patterns. She bunched and contracted her muscles under my hands – and weed all over me. She also gave me a heavy dosing of pungent yellow musk from the vent (bottom), which a lot of large snakes do when they're handled. After catching anacondas a couple of years ago, I had to throw all my clothes away – it was worse than being zapped by a skunk!

New Guinea's herpetofauna has much in common with that of Australia, including many of its most venomous snake species, and it is also one of the few places in the world where venomous species outnumber the non-venomous ones. The taipans are the Australasian equivalent of the black mamba – long, quick, angry in defence, and delivering bites containing enormous amounts of virulent nerve toxins that can penetrate deep into the muscles of their mammal prey.

Another potential nasty that occurs here, and is actually responsible for far more deadly envenomations than the taipan, is the death adder.

In the rest of the world, the word 'adder' is synonymous with viper. Vipers are usually quite short snakes, with thick heavy bodies and large, arrow-shaped heads. They sit camouflaged for long periods waiting for prey to come close, and then strike rapidly, often releasing the victim to die from the effects of the venom. A viper's fangs are hinged, so they can swing forward to stab into their prey, and the main constituents of their venom are haemotoxic, which means they affect the circulatory system. The death adder looks and behaves just like all the other vipers – but it isn't a viper at all. In fact, like all of Australia's and New Guinea's venomous snakes it's an elapid, or member of the cobra family.

This is an instance of convergent evolution, when two animals in different parts of the world evolve to fill a similar niche and thus end up looking similar. The first example of this phenomenon usually given in biology textbooks is the wing, which appears to have evolved many times in history and in many groups of animals but always has similar properties, as the wing structure's evolution is driven by the laws of aerodynamics. In the same way, the extinct thylacine or Tasmanian tiger mentioned earlier bears undoubted similarities to placental canids or conventional dogs, but is actually a marsupial. The egg-laying mammal the echidna has evolved prickles like those of hedgehogs, porcupines and Madagascar tenrecs, although these facets evolved in each animal separately.

Vipers are the snakes that cause the most human fatalities. Around twenty thousand deaths a year are attributed to the shoelace-length saw-scaled viper in the Indian subcontinent alone. The simple reason for this is that their hunting strategy makes them more likely to take up position alongside tracks and trails, and to stay there and not slither away at the approach of footsteps. They get trodden on, then bite back in self-defence, leaving their human victims in big trouble. Despite being a member of the cobra family, the death adder also stays still until you're unfortunate enough to step on it. Its venom being viciously potent, it was the creature most threatening to our safety on the expedition, so when one of the New Guinean trackers, Maxi, mentioned he'd seen one along the trail not so far away, I set out to make sure it wasn't left there, where it might cause trouble.

I took to Maxi instantly. He walked barefoot through the jungle in scrappy old cast-off shorts, decorated his beard with little orchids and wore a biro pushed into his tight curls. He had an infectious laugh, which took me by surprise. Despite being muscularly built, with a serious demeanour and thick beard, and looking as though his laugh should be a deep baritone, he would let rip with a high-pitched hee-hee-hee-hee, which sounded more like the giggles of an eight-year-old schoolgirl. He was kind, hard-working and intrigued by everything we did – and wonderful company, despite his mix of English and pidgin being sometimes difficult to interpret.

Maxi and I walked for about half an hour out through the forest to the spot he'd described, 'a junction between two paths, twenty yards after the camera trap that's been set up to try and film megapode birds, under the big dipterocarp tree'. When we got to the spot, I told Maxi not to show me where the death adder was, so that I could find it for myself. He could barely contain himself, giggling away with his hand over his mouth as I wandered clueless amongst the leaves, clearly almost stepping on the snake several times. He kept hugging himself in an attempt to contain his glee, jumping up and down with mirth as I leapt six feet in the air after stepping on the end of a stick, causing the other end to twitch in the leaf litter.

Finally, after about ten minutes I noticed the tip of the lethal snake's tail flickering in the leaves. The tail tip is bright yellow, and this 'caudal luring' is done to attract small mammals, birds and frogs that might mistake the tail for a wriggling caterpillar meal.

'*So you pela savay im snek?*' Maxi asked me. 'So you're the guy who's supposed to know about snakes?' He didn't look very convinced.

Once the death adder was lifted into the open it seemed insane that it could be so hard to find. It was a fat snake, not much over a foot long, but as thick as my wrist and with a lurid striped pattern down its back in bright orange, red and black. Place it on a patch of grass and it'd stand out as much as a prom queen in a prison. However, the second I took it off the path and placed it far from passing feet and back into the leaf litter, it disappeared completely into the motley of colours.

'*You step on im snek, is dai pinis,*' Maxi pronounced 'If you trod on that snake, it'd be a painful death.'

So while snakes were my department, George took the insects and Allen the frogs, Jack Dumbacher was our bird aficionado. I was in awe of Jack even before we got here, as I'd read one of his papers on the wonders of the pitohui, the world's only poisonous bird.

Way back in the early nineties, Jack and his team had noticed that certain birds were shunned by the New Guinean locals. They called them 'rubbish birds' and refused to eat them – quite a thing amongst hunters who would seemingly eat anything the forests provided. Jack also noted that when pitohui were caught in mist nets, the researcher handling them would often end up having a sneezing or coughing fit. If that same researcher later touched his lips or other sensitive bits of skin, he'd get tingling sensations or fierce itching. When Jack ran tests on the feathers and pelts of the pitohui, he found they contained batrachotoxin, which is the same kind of poison found in poison-dart frogs. These are the most toxic creatures on the planet, one Colombian species being the size of a fingernail but containing enough poison to kill ten men. Like the dart frogs, it seems pitohui birds sequester their toxins from invertebrate prey. In this case the prey was probably choresine beetles, which in turn get their poisons from plant alkaloids. The poison is used by the pitohui either to ward off parasites, or to deter predators.

Like Paul Shearman, Jack is an extremely modest man, but with knowledge and experience that makes me feel truly small. Every day Jack brings in some new miracle bird he has caught in his mist nets: several different species of pitohui, a kingfisher with azure throat feathers, a beautiful fruit dove that sits in my hand, its tiny heart fluttering through its scarlet and jade plumage. It could well be the most aptly named bird in the world. And then the treasure, a king bird of paradise. I remember the first time I read about birds of paradise as a wide-eyed boy. When other kids dreamt of playing for United and kissing Michelle Taylor behind the bike sheds, I dreamt of finding a bird of paradise in the jungles of New Guinea. And here it was, the king bird of paradise, a perfect crimson miracle the size of a starling, the wires from its tail dropping shimmering eye discs like the eye spots on a peacock's tail. It is the smallest of the birds of paradise, but more than makes up for its diminutive size with its dazzling colours and displays. When the male is showing off for a female, he fluffs his tail feathers out into a massive puffball skirt, then swings upside down on a branch flashing his shimmering discs. It was utterly sublime, one of the most beautiful things I have ever seen.

The evolution of the bird of paradise's phantasmagoric feathers is a subject almost as fascinating as the birds themselves. Only the males are thus adorned – the females look like nondescript crows. It's likely that the colours began to evolve because of the dark forests. The showier and more colourful males were more likely to stand out, and therefore to find a mate. These successful males passed on their genes to descendants, who became more and more colourful through the generations. However, there has to come a point where being easily seen is a disadvantage, and those huge twirly tail wires and hefty feathers definitely make it harder for these birds to fly. Both of these features have to put a bird at more risk of being spotted and then caught by predators, so why did they continue to develop?

A solution is elegantly provided by the handicap principle, posited by the Israeli biologist Amotz Zahavi in the mid-1970s. The handicap principle states that certain animals develop showy mechanisms that seem to make them less fit for survival, but make them more attractive to mates. The peacock's mammoth tail is the first example that's always given. Effectively, what this tail says to peahens about the peacock is: 'I'm *so* incredibly strong, fit and healthy, that I have survived to breeding age despite the fact that I'm carrying around a barrowload of brightly coloured feathers. I don't need to hide, I can stand out in the open, and nothing can catch me. Just imagine how strong I must be!'

This theory makes sense of so much of the bird of paradise's behaviour as well as of its colours. Not only the unwieldy feathers but also the courtship dances and songs practically scream out to hunters to come and catch the bird. Jared Diamond suggests that certain human showing-off behaviours, like driving sports cars too fast, getting incredibly drunk, bungee-jumping or playing contact sports, might be evidence of a form of handicap principle. Essentially it's men saying: 'I'm so amazingly fit that I can do really stupid stuff that endangers my life, yet survive to be a successful breeding partner!' Perhaps our posturing here in the jungle was just more of the same . . .

No more than a kilometre or so upriver from camp, a large, low sandbar with thick reeds behind it juts out from the bank. Early one morning we were cruising past the sandbar, and from the boat I spotted tracks leading up through the sand. We pulled in close to shore, wading the last few metres so as not to spoil the tracks and destroy the message the creature had left for us. Spidery clawprints led out of the water, with a dragging, swaying tail track between them. The size of the digits, the distance between the feet across the breadth of the animal, and the stride length, lead me to deduce that this crocodile was probably two metres long and slightly built for its length. The fact that the tracks led in a neat circle up out of the water, curving round towards the back of the beach then straight back into the swell, without any sign of having stopped to bask, made me think this lady was patrolling here, coming up out of the water to check on something before returning to the sanctuary of the river.

We looked around us nervously. She could well have been watching us, but the churning river was the colour of milk chocolate, so we'd never know. The intriguing thing would be to find out what she was guarding – was there a nest nearby? With my heart thumping with excitement, we followed the direction of the tracks up into the forest at the edge of the river. Everyone put down their cameras and sound gear and we employed a search pattern, criss-crossing the brush. A crocodile's nest is no more than a scraping together of the available leaf litter, earth or sand. They can nest at the water's edge, or quite deep in the forest. An experienced crocodile mother will judge well and nest above the highest flood or tide mark, but that's not always the case. Crocs come back to their nest to protect it from interlopers, and once the hatchlings have used the egg tooth on the end of their snout to cut their way out of their leathery-shelled prison, the mother will come back and carry them away gently in her mouth. That parental care can last for months.

The crocodiles in New Guinea are some of the largest in the world. The salties or estuarine crocodiles can live easily in salt or fresh water, and are found both out to sea over coral reefs and way inland, like here. Big males have been known to exceed six metres in length and weigh over a tonne. An animal of this size has no predators other than man, and can bring down a fully grown Asian water buffalo; and even sharks, when they hunt out at sea. One or two people a year are taken by crocodiles in Australia, mostly individuals who have drunk too much and either fallen asleep at the riverside or decided to go swimming in areas that are known croc haunts. In other words, even in areas where crocs are numerous, attacks are avoidable, with a little common sense. Mind you, big crocs that may be struggling to hunt naturally will certainly grab a human meal if it presents itself.

On a recent mission in the Northern Territories to catch and relocate an injured four-and-a-half-metre croc known as 'Old Stinky', we spotted his night-time eyeshine lying just metres below a fisherman who was standing on the riverbank, totally unaware of the monster lurking just below him. On New Guinea's Sepik River, some large crocodiles have terrorised villages. There's a story of one leviathan croc taking thirty people before being caught. Due to persecution and hunting for their skins, these huge specimens are very scarce nowadays.

Back to the sandbar. After about twenty minutes there was a shout from cameraman Ian, working on his first expedition series and making a good impression with his hard work ethic: 'Steve, I've got the nest!'

It was on elevated ground between two trees. No one would ever have noticed it, but the top of the nest had been scraped open and three eggshells lay there, torn apart.

'It looks like they've hatched out,' was Ian's first thought, but I could instantly see this wasn't the case.

'No, I don't think so, mate. First up all the eggs in a clutch would hatch in the same night, but look down here.' There were still at least seven or eight eggs left in the nest. 'There's a little trick as well – just hold that complete egg there up to the sunlight.'

The egg is goose egg-sized, though longer and thinner, and the shell is leathery-soft, not brittle-hard. Ian holds the egg up.

I explain further: 'The shells are vaguely translucent. If you hold a late-term egg up to the light, you can see the little croc inside, sometimes you can even see them wriggling and reacting to torchlight.'

These, however, were all quite early in their development, so clearly not about to emerge.

'There's one more thing,' I continue. 'Hatchling crocs have an egg tooth that cuts through the shell and leaves a clean exit, with very little amniotic fluid left behind, but these are all shredded and messy. No, this nest has been raided, something's dug up the eggs and eaten them.'

'So what do you think?' Ian asked. 'D'you think some of the locals' dogs or pigs have been through here?'

That was an interesting idea. I hadn't thought of something as simple as that. No, my idea was that this was the work of *Varanus salvator*, the water monitor lizard. There was only one way to find out for sure.

'Whoever raided this nest is gonna be back this evening to finish off the job.'

Our best bet was to frame up an infrared remote camera and see who paid a visit. Ian placed a camera trap pointing down at the nest, while I half-heartedly buried what remained of the eggs, knowing this wouldn't make any difference.

The next morning Ian and I returned to the croc nest, and found the scene just as we expected. On the beach was another set of prints from our distraught crocodile mother, but she had been unable to prevent her nest being devastated. There was nothing left but a few sad pieces of shell.

Too excited to wait until we could return to basecamp, we took out the remote camera and scanned back through the images. There were a couple of shots of Ian and me looking into the lens as we came in to pick up the camera, then *bingo*! Framed beautifully in the centre of the next shot was our culprit, crocodile yolk dripping from its dinosaur maw. The water monitor lizard is the second-largest lizard in the world, and looks almost identical to its close cousin the Komodo dragon, which is the largest species of monitor. It has all the latter's lethal attributes – the fierce curved teeth, slavering forked tongue, talons like an eagle and a whipping tail much like those of the crocs they steal from. The water monitor is a bold, ballsy, hypersensitive predator and scavenger.

Several weeks later, when I am out of camp, yells come up from the croc beach as all hell breaks loose and George mobilises one of the outboards. Somehow, communications have broken down with the local guys who have been driving the boats for us. They strung one of their fishing nets right across croc beach, and caught considerably more than they bargained for. Our female crocodile has swum into the net, become entangled, and is thrashing around like a beast possessed. Within a very short time she exhausted herself and drowned. Had this taken place in one of their villages, the locals would take the crocodile as a blessed gift of extra protein, but as we are making a conservation series and she is one of our stars, George shouts at them to put down their machetes. Instead, they have to undertake the far more dangerous and difficult task of cutting loose a thrashing, hackle-raising horror of teeth and thrashing tail. George dives in to help out, cutting the net with his penknife as the monster is unleashed. Later on, when I hear what has happened, I'm thoroughly frustrated not to have been there. When I watch the footage later and see the crocodile they've caught, I detect a very definite feeling of smugness. 'Yes, she's about two metres long, but slender for her length.'

Dr George McGavin, who heads up the science team, is something of a talisman for the whole team. George is the lifeblood of the camp, a larger-than-life Scotsman who has been trapped for four decades in a university too small for his personality. Now he's like a teenage girl let loose with a mobile phone and infinite credit.

After our first expedition together in Borneo, George offered me the privilege of going to work for him at the Museum of Natural History in Oxford. It's probably my favourite building in the world, an imposing but not overwhelming Gothic structure with divine high-arched ceilings, but it's the dusty little oddities in the cases around the place that I find really enchanting. Dinosaur skeletons sit alongside beasts that were stuffed and shipped home by the Victorians in those golden days of exploration. Out in the back is the Pitt Rivers anthropological section of the museum. Half a million ethnographic curios dwell here, including shrunken heads and even a witch who's been constricted in a tiny bottle for several hundred years, and no one dares open it.

I relished my time there with George, relabelling the preserved insect specimens in the Hope Collection, rooms full of huge old camphorwood and mahogany cabinets, each with rows of glass-topped cases containing hundreds of thousands of insects from all over the world. There were specimens from the 1700s, the oldest in existence, and some that'd been preserved and donated by Darwin himself. One of the huge storerooms was once a lecture hall where Darwin gave his first public announcements on evolution. Apparently the captain of his ship the *Beagle*, a fiercely Christian man, heckled loudly from the back, heart-broken and horrified that his ship and all his efforts had given rise to such heresy!

The short time I worked there, with its mothball-scented corridors and redolent with such a depth of history, was an immensely exciting time for me. It gave me a taste of the dusty academia that I'd missed in my own education. But George, after decades trapped there, was just bursting to get out and put his vast knowledge into practice. Now, every day in the jungle he produces weird weevils with giraffe necks or neon-blue wingcases, jumping spiders with palps like giant boxing gloves, stick insects as long as your forearm and moths like flapping paperback novels. He cheerfully spends an entire day with a jemmy, breaking apart a rotting log in search of beetle grubs. Covered in rainforest grime, he's as happy as a pig in poo.

George's particular quirk is a unique ability for self-inflicted blood loss. In the space of a day he manages at one point to stand up enthusiastically in a cave, declaring: 'Wow! The roof in here is really' – *thwack* – 'low.' Then, walking down a muddy slope he warns: 'Watch where you put your feet, guys, this stuff is really' – *schwoop . . . thwack* – 'slippery.' Next, he nearly tears his ear off on a thread of rattan, a jungle creeper as strong as nylon rope and covered with pairs of evil curved claws. It uses those claws to cling to other vegetation and catch a lift skywards towards the sunlight, and is one of the world's most vicious plants. If you catch on it and try and tug away, your flesh will give way long before the rattan will. This plant is in many parts of the world known as 'wait-a-while', and is behind the jungle mantra, 'You can't fight the forest'.

One of George's schemes is to set a moth pupa he has found in a makeshift darkroom surrounded with time-lapse cameras in the hope of seeing it emerge. Ulla, the camp photographer, sleeps on a bench by the darkroom, getting up every few hours to change the batteries and memory cards. For four days. And then one morning George looks inside the darkroom and declares: 'Oh, it's gone.' The camera crews leap into action to film the denouement of this miraculous event. We replay the footage with everyone looking on in anticipation. After a few minutes, an innocuous brown slug-like shape is glimpsed crawling out of shot. The whole thing has lasted for a single frame. Ulla attempts to put a positive spin on things while trying not to cry. Everyone else holds in their chuckles until we've retired to a respectable distance, then, a bit sadistically, we laugh quite a lot.

Luckily, we are not putting all our wildlife-filming eggs in one basket. We have brought along some of the best film-makers in the world. Gordon Buchanan left his job washing dishes in a canteen on his native Isle of Mull at seventeen to take up an apprenticeship with Nick Gordon, one of the best-known legends of natural history camerawork. Gordon worked as his assistant cameraman for several years, living with him and his menagerie in the Amazon.

As you'd expect from someone who learnt their trade from the very best, Gordie has become world-renowned. In addition, he is annoyingly attractive to women, with striking features and distinctive curly white hair. He speaks in a Scottish brogue that could be put to good use doing voice-overs for the Scottish Tourist Board – 'Visit Scotland, cos there's such a thing as too much sunshine!' – and has a laconic, self-deprecating manner. Gordie says it's impossible to be any other way, living somewhere as small as Mull. Some of our hit TV shows have had millions of viewers and garnered prestigious award nominations, but he'll go home to Mull and they'll greet him with: 'Ay . . . saw your show the other night – wha wa tha rubbish?' The one thing that stops us all hating Gordon's guts as Mr Perfect is that, in true Scottish style, he's so pale he's almost blue, and burns like a vampire at the mere suggestion of sunlight.

At night in camp the two Scots, George and Gordon, are the main source of hilarity, making a great double act. George's favourite trick is to approach someone who knows nothing of his irreverent character and to whom he is the illustrious 'Professor George McGavin, eminent entomologist from Oxford University'. Very seriously, he'll offer them his extended forefinger and ask them to pull on it, as if inviting them to take part in some important scientific experiment. When the respectful but bewildered subject does as asked, George will let fly a generous and reverberating fart.

One of his favourite tales is of when, researching human lice, he consulted some Victorian textbooks on the subject to try and get some perspective. He found the following passage: 'The pubic lice live in the hair that surrounds the genitalia of adults. Sometimes it be found even in the beard and eyebrows, though it is a mystery how it arrives there, as it crawls not, neither can it fly.' His impression of the mystified upper-class Victorian who just cannot conceive of how pubic lice could get from one lady's garden to a gentleman's moustache has us all in hysterics.

Gordon tells us of the results of his day. He has been trying to film pygmy parrots, the world's smallest parrot species. The filming has not been going well, as these birds are about the size of your thumb and frequent the darkest, dingiest corners of the forest. However, a stroke of luck presents itself when Jack finds one entangled in his mist nets. He brings it back to camp to take some genetic samples, and so that we can film it. Of course, this is an instant focus for the cameras, which gather like flies around a corpse. Lighting rigs are assembled, microphones placed on to Gordon and Jack, and filming begins. Gordon takes the precious bird in his fingers and starts to tell its story, while Jack prepares to run his tests.

'It actually has quite a nip for such a tiny bird,' Gordie begins. 'In the forests their feathers – oh, ow, OW!', as the miniature parrot takes a chunk out of his thumb. 'Oh shit no!' Unforgivably, Gordon has let the parrot go in a flurry of tiny green feathers. His face is a picture of horror – before he collapses into giggles.

'But it's OK,' says Jack, ever the professional, 'because I needed some blood and he left behind some feathers, so I can run my tests on those.'

Gordon is still in shot, trying with all his might to control his near-hysterics. That evening at dinner, he lets us in on his own personal theory of New Guinean symbiosis: 'You just know that with the way things work round here, if there's a pygmy parrot, that somewhere in this forest is a miniature pirate walking around . . .'

Three weeks in, and even the oasis that is basecamp began to seem like a waiting room for the condemned. One of the facets of camp life that even the battle-hardened struggle to abide is long-drop toilets. The long drop is a three-metre-deep pit with boards over the top that you sit on for your essential functions. When you lift back the lid so as to sit down, a swarm of flies, wasps and other nasties erupt from within and make a bolt for freedom through the very hole you're about to sit on. People in camp basically fall into two categories: lookers and non-lookers. Those who go out at night and do anything they can to avoid letting their headtorch light penetrate down into the nightmare of seething white maggots in the excrement beneath, and those who cannot stop themselves from having a look – a gaze into Hades itself.

To the astonishment of everyone, one of the camera crew admitted he took to the toilets when it all got a bit too much and he needed some 'me time'! How desperate would you have to get to escape to the long drops to be alone?! One night the horror got even worse when a substantial flood rushed through camp. The slurry flushed several weeks' worth of shit and maggots out of the long drop and up to the surface. Two days later, a plague of flies arrived. We lit fires around camp to try and keep the winged horrors at bay, and sat choking in the smoke to try and avoid them. Then someone came up with the bright idea of painting the table-tops in the food tent with garlic oil in an attempt to repel them. This just led to everyone's clothes and skin stinking of rancid old garlic.

One particularly miserable evening, to ease the pain Tim Fogg produced a small bar of Green & Black's chocolate. Eyes glint wolfishly in the dark as he deals out the precious taste of civilisation. One of the locals takes more than his share and normally easy voices take on a steely edge. Ulla and Gordon take turns nibbling on a single square, like dormice round the last hazelnut in the world. As Jonny Keeling walks out with the last square clutched in his hand, he slips on the greasy boardwalks and nearly backflips, coming down with a sickening crunch on his muscular back. The impact is so severe that his metal canister of bug spray is squished flat, but he leaps up and dusts himself down, the archetypal alpha male who'd sooner die than be seen to be hurt. But then there's utter panic – where's the chocolate? For several minutes six hardened biologists squint with their headtorches in the mud, looking for a single square of precious sweetness. It is found, brushed down, and devoured.

That night, my Irish cameraman and dear friend Johnny Rogers came into the kitchen area ashen-faced and took George aside. After a little furtive chatting George started roaring with laughter, then went over to the lab, got his entomologist's magnifying spectacles – huge rectangular lenses that fit a few inches in front of his normal spectacles to give him a more intimate view of the bugs he studies – and passed Johnny a pair of long tweezers. He took them as if they were highly illicit cargo, and headed off to the toilet. Half an hour later he returned rather sheepishly.

'One of the girls just walked in on me Stevo,' he confided.

'How's that, Johnny?'

'Well I was in the long drop you know . . .' He tails off.

'God, don't worry about that Jonboy, everyone's scummy as hell here, nobody's going to be offended by walking in on you on the toilet!'

'NO! It's worse than that! You see I found three ticks in the end of the old boy and on the cooblers!' (In Johnny's Northern Irish patois old boy and cooblers are the meat and two veg.) 'The ticks were just tiny wee ones ya know, I could barely only just see them, so I took George's bug specs and some tweezers and—' He mimes pulling his penis out tight in front of him and scrutinising it through the specs.

'Not surprised Jonboy – I guess you'd need some serious magnification to find your old fella.'

13

The world downriver is steeped in mystery, totally unknown to outsiders. Chinese whispers at basecamp tell this to be a mission too far, an expedition that could produce our first fatality. Knowing the malevolent nature of our river, even hardened adventurers quake at the idea of abandoning themselves to the limestone gorges. Once within their vertical cliffs there would be no escape. Gigantic waterfalls plunge down into the river from either side, more falls than I've ever seen in one place, creating vast spray clouds that billow upwards, sparkling in the sunlight.

While our team covers the miles around basecamp metre by metre, soon it's beginning to feel as if we're retracing our steps. I'm itching to get out and explore further. Our helicopters have been flying recces in the southern highlands, and have discovered about twenty miles downstream a narrow canyon lined with extraordinary waterfalls. To our knowledge no one has been down there before, and to me it seems to be a particularly exciting destination. Anywhere that is totally unknown, and where our footsteps are bound to be the first, is to me the most enticing of prospects.

I work myself into a fervour, imagining the wonders we will discover in the world downstream. But when we gather the team together to start talking logistics, I find that my excitement is not shared by everyone. In fact, some of the team genuinely think that to take the RIBs (rigid inflatable boats) downriver would be effectively committing suicide. The Hegigio River which runs outside our camp is not to be messed with. It's as wide as the Thames at Windsor or Marlow, but runs with the force of a cataract.

In terrain where walking a mile can take a day, a river is the most valuable asset. It's a potential freeway, which can allow quick and direct access to places that would otherwise be beyond your grasp.

In order to use the river to its full potential, Steve has imported two inflatable Zodiac RIBs with powerful rear-mounted engines. From my point of view it's all a bit noisy, not to mention too easy to travel by motorised transport rather than human power, but the ferocity of the Hegigio makes this our only option. The team gather in the kitchen tent and stand around a table with maps and some photographs that've been taken from the helicopter.

I take the lead. 'Right, well, we've penetrated about six miles downriver from here with the Zodiacs, and we all know what this river can be like. It's full, and running with a lot of volume, but there's not much in the way of rapids. Further than that nobody has to our knowledge ever been. Perhaps thirty miles downriver there's another village, probably of a hundred-odd people – here.' I point at the map. 'I seriously doubt we'd need to portage at all' – portage is the term for getting the boats out of the water and carrying them around an obstacle – 'but just in case we do, I'd push for going with a small team and the absolute minimum of kit.'

'What worries me,' says safety guru Tim Fogg, 'is what happens if one of these Zodiacs conks out. This section here' – he points to the most exciting-looking area – 'is a steep-sided gorge – there's no saying we'll be able to get out anywhere here.' He points to where the topographical map shows hundred-metre cliffs dropping in a squeeze of contour lines, evidence of vertical terrain. 'If the engine dies, you'll just get flushed all the way down to here.' He points to where the terrain levels out. It's a long way.

'This is just stupid,' says Johnny Rogers. 'At what stage are we going to say enough is enough? I've got kids, you know, it's only telly.'

Nick the soundman agrees: 'We have no idea what we're getting into here, boys. This could be a step too far. If we just keep pushing it something's eventually going to go bad.'

'Come on, Johnny!' I was incensed. 'This is nothing like as dodgy as Kuli' – the mountain we climbed together in Borneo. 'We're in motorised Zodiacs, and this river's sketchy but it's never going to be more than a class three.'

The last thing I want to do is push anyone into anything that might risk their lives, but it just seems that everyone's winding themselves up into an unwarranted frenzy of fear.

'Absolutely worst case scenario,' I continue, 'the engines die and we drift downstream. There's not going to be anything gnarly enough to tip you out of the boat. We have the second boat to effect a rescue, and if that goes too we float down to the village and sit it out until the cavalry can come get us.'

'Nobody here knows how to run those boats, man,' says Nick. 'If we had decent drivers then I'd say no worries.'

'No way I'm putting my life on the line with you driving the boat, Stevo,' Johnny adds. This stings, but is a totally fair call. I have a real aversion to motorised transport of any kind, have been far from willing to practise in the Zodiacs, and as a consequence am not massively confident at driving them.

'We can't rely on any of the locals either, man, most of them can't even swim,' Nick says. 'They're shit-scared of the water, and have no idea about controlling the RIBs.'

'Nick's pretty good,' Tim counters. He's talking about one of the local guys who is also, confusingly, called Nick. He has only one eye and a piratical air about him, but has been driving the boats with enthusiasm and confidence since our arrival. He seems trustworthy to me.

'He's never done anything like this, though. It'll take three days to get down there.'

'Three days?!?! You must be mad, Johnny!' I try not to scream with frustration. 'Five hours tops – it's only about twenty or thirty miles.'

'You're dreaming, Stevo.' Johnny can be fabulously stubborn sometimes; he has his mind made up and is not for turning.

'I don't think it's going to be that bad,' says Tim. This is good. Tim is the calming voice of God when it comes to safety on the expeditions. 'From what we've seen of the river it looks highly navigable, and it's always wide enough that we can skirt around danger.'

'No way, Timmy,' says Johnny. 'It's just too much that we don't know about. I'm not prepared to leave my kids without a dad for telly. I'm out.'

I was gobsmacked. Johnny had been my fondest constant companion on the expeditions, dragging himself up unclimbed peaks, down waterfalls and into unexplored caverns with me, always ready with a cheery word and a nip of Irish whiskey at the end of the day. He's always managed to keep filming, missed almost nothing, and been one of the vital ingredients to the success of some of the gnarliest things we've done. Additionally, we have a bond that goes deeper than just presenter and cameraman. One time, on the mountain climb in Borneo, on a near-vertical slope, he grabbed a root with both hands. He put all his weight on it, and it came clean out in his hand. For a millisecond there was a moment of stunning clarity, like Wile E. Coyote at that second when he runs out into space off a cliffside, stops in mid-air, and realises he's about to fall.

Johnny was about to drop to his death maybe a hundred metres below. I saw the trees way down the mountainside aimed at his backside like spears, and had a mental image of myself breaking the news of his death to his wife, all in less than a second. How I reached out, grabbed his arm and managed to pull him back to safety, though, I have no idea at all. Then, just minutes later, Johnny repeated the favour when exactly the same thing happened to me.

When someone's saved your life and you've saved theirs, there's a special bond that will always be there in the back of your mind. We owe and trust each other. I wanted Johnny along.

'Come on, Johnny, I don't want to put pressure on you, but you're being a total pussy!'

'No way, Stevo, this one's too risky,' he says. 'Anything could happen down there and there's just no plan for getting us out. It's just a job, it's not worth dying for.'

I look at the maps again – is there something here I've underestimated or overlooked? Am I missing something? I just don't get it.

'I'm willing to give it a go,' says Nick, 'but we have to be careful, man.'

'But who's going to film it?' I ask forlornly.

'There's always Lucky Luke,' suggests Steve Greenwood.

Luke is a South African cameraman, enthusiastic, fit and good company. I don't know him well, but this seems like our only option. Steve is doing his General Custer impression, hands pressed into the small of his back. He looks very worried indeed. We all look to Tim. Ultimately it will be his call. Does he consider it a gamble worth taking? He looks a little nervous, which is enough to set the butterflies going in my stomach. However, he nods cautiously. We're on.

As if to prove the doubters right, that very afternoon I was running the Zodiac out from a small creek on to the main river, and the engine stalled and stopped dead. As I frantically tried to restart it, yanking on the starter cord, our little boat got caught up in a furious whirlpool at the mouth of the creek, and all hell broke loose. It got spun round and round in dizzying circles, and on each turn the side was dragged under water and the brown river poured in. Pretty soon we were up to our knees. It was clear that the engine had flooded and would not be starting, and that we were in for a very rough ride.

I abandoned the engine and took up my paddle, frenziedly trying to pull us out of the maelstrom, as the folks in the other boat yelled and screamed at us from outside the whirlpool. Rescue for the moment was impossible. Finally, the rapid spat us out and we were cast off downstream, dragged under a rocky overhang, then out into the rapids again. Relieved, I breathed out and flopped down in the boat, but the other boat was yelling again: 'PADDLE, BLOODY PADDLE, STEVE!' I looked up, saw that we were passing the safety of a rocky bank, and just beyond it downstream was a fierce rapid that could spell danger. I struck for shore, and at the last moment jumped out into the flow and dragged the boat to safety.

The whirlpool escapade had its positives, though. As we were emptying out the boat, I turned over a stone to find a giant scolopendra centipede as long as a pencil and three times as thick. These creatures have a viciously venomous bite, delivered by two modified claws at the front of the head. It's said to be one of the most painful invertebrate bites in the world, but I've learnt through experience that you can allow them to scuttle over your hands safely as long as they don't feel restrained or threatened. I duly show this off to the camera. However, what I hadn't figured on is that centipedes like to have all surfaces of their body enclosed and seek out the security of dark, tight places. This great venomous centipede did a brief skitter over my arm before finding my watch, then slid straight under the strap and curled up tight, tangled through the buckle against my skin. This led to something of a dilemma, and it took me about five careful minutes to get my watch off.

With the thought hanging over us that this is to be a daunting mission, likely to be at some stage life-threatening, the preparation is intense. Every tiny item of kit is scrutinised and every scenario run through a hundred times. The boats are loaded with throw lines for rescuing anyone who goes overboard, engine spares are collected and the motors tuned up. Everything that will come with us is painstakingly stowed in dry bags.

The mood in camp is quite sombre as we continue packing for our journey downstream. The foreboding of the doubters has spread, and some of the crew honestly think they may never see us again. For a while, the tension infects me too, and I find myself anticipating the worst. Finally, early in the morning of departure day, we say our goodbyes. There are hugs and good wishes all round. It feels as if we're soldiers heading off to the trenches. When we're all aboard, I sit at the front of the boat and deliver a very worried piece to camera to the tune of 'We are heading to our possible doom ... once we're into that canyon we're on our own.'

Somewhat inevitably, we head off downriver prepared for a hellride we might never return from. In the event, it turned into a fabulous cruise through one of the most beautiful prehistoric environments I'd ever seen. Maybe six miles downstream of basecamp was as far as anyone had ventured before, so we immediately felt as if we were blazing a trail into the wilderness. The shallow forested slopes prevalent around basecamp soon disappeared, as the river plunged into a gorge with vertical limestone cliff faces a hundred metres high on both sides. Massive waterfalls plunged down into the river every hundred metres or so, creating vast clouds of spray billowing up into the sunlight.

I desperately wanted to write a whimsical piece telling how this place had been unchanged since T. rex ... Unfortunately, this would have been out by about sixty-four million years. In fact the geology here is practically newborn, little more than a million years old. Every one of the waterfalls is spectacular enough that if it were in Europe people would travel hundreds of miles to see it. Here in the forests of New Guinea there are hundreds, sometimes four or five down one riverside in the space of a couple of hundred metres. It is one of the most awe-inspiring places, truly a natural wonder of the world. The promised terror ride, though, couldn't have been less in evidence. All our lifelines, lifejackets and secured equipment looked totally over the top as we roared through that paradise on our outboard-driven dinghies. I'd had scarier drives down to Tesco.

Finally, after a long first day's drive, we pulled up to the riverbank opposite two huge falls, and set to making camp on the flat floodplain above the water. It was one of the most stunning places I'd ever camped in, marred only by the presence of huge tabanid flies. They seemed to actively enjoy the taste of chemical bug spray, but undoubtedly loved the taste of blood even better. Like mosquitoes, the male tabanid flies are nectar-lickers – it's the females that feast on blood to garner essential protein to nourish their eggs. However, while a female mosquito's mouthparts are akin to a syringe for sucking the blood through, tabanid flies have scalpel- or steak knife-like mouthparts which slice open the flesh. Then they lap up the blood with tongue-like structures. The bites hurt like hellfire, and result in huge itchy swellings. As we began rigging up our hammocks and cooking up some food, we swung our arms about our bodies all the while to stop the flies getting a purchase long enough to bite.

That night we donned headtorches and trekked around the forest near the camp in search of nocturnal wildlife. The most numerous and diverse creatures were without doubt the stick insects, which decorated the trees in a befuddlement of colours and different body forms. Stick insects belong to a group called phasmids, which means 'phantoms' and refers to their ability to disappear spirit-like into their background. To those whose only experience of them is the spindly green Indian stick insects so beloved of school science labs, these wonders would be truly mind-blowing. Some are as long as a human forearm, with great Chinese-umbrella wings kept folded by their sides. Others curl their segmented abdomens up over their heads in a poise that must surely be mimicking a scorpion. Some of the great heavy armoured sticks have spined legs with which they can kick their tormentors, easily drawing blood from human fingers, for instance. They are endlessly fascinating.

My jewel of the evening, though, was a tiny web-weaving spider in extravagant colours, feasting on a cricket that had witlessly sprung into her web. Spiders are the single greatest worldwide cause of insect mortality, being responsible for an estimated 99 per cent of all grisly insect deaths and eating in bugs the equivalent weight of the human population of England every year! That night's arachnid splendour was no bigger than my little fingernail, and it had an abdomen shaped like the pope's mitre, in bright scarlet and aquamarine. She looked as if she'd been embellished with metallic paints by a perfectionist model-soldier enthusiast, and she'd made a perfect web glistening with late-dusk rain droplets.

As I worked my way along the floodplain, my headtorch suddenly caught a bizarre sight. Hung upside down from a tree branch was one of the oddest-looking bats I'd seen. It was a warm orange colour, about the size and furriness of a kitten, with its wings wrapped around it like Dracula's cloak. It appeared to have a tumour the size of a fist growing on its face. As I got closer, though, I could see that we had merely disturbed him while he was tucking into a fat, juicy fig. He couldn't make up his mind whether to drop it and fly off, or just sit there quietly. It was as if he was saying to himself, 'Just sit still, and maybe they won't notice you.' Rather like a stunned, wide-eyed kid who'd been caught in the act of trying to devour a chocolate beach ball. To add to the weirdness, this was a tube-nosed blossom bat, and in place of a neat fox-like fruit-bat's nose he had two long tubes curling out of the end of his snout – almost as if two fat earthworms were crawling out of his nostrils.

The next day, we decided to scale one of the waterfalls opposite, to try and explore the world at the ridge above. We drove the RIBs over and roped them up at the bottom of the fall, which at that point was about thirty metres across. The water tumbled over multiple steps to the river, but we couldn't see the top as it was engulfed in forest. Though the fall was constantly flushed through with water, the limestone underneath had remained spiky and grippy and could be scrambled up. The water rushed into our faces as we climbed. It took the best part of the day to reach the top, scrambling to the side when it got too steep and hacking through the impossibly thick, spiky undergrowth. At the top, though, disappointment awaited. The last pitch of the climb was vertical rock, green, crumbly, overgrown and with no chance of circumventing it. It would have been an edgy but achievable climb with ropes, but we didn't have any, and the thought of what would happen if anyone got injured up here was just too hideous to contemplate. Disconsolate, we turned tail and started to descend.

The next day we began our journey back upriver to basecamp, despite not wanting the divine spectacle of this magnificent place to ever end.

But on our return we had a very close call and a few moments of genuine panic, probably due a little to tiredness, but also to complacency. We let our guard down briefly, and the results were nearly quite nasty.

We were powering upriver through a wave train behind a rapid, and were trying to keep the boats side by side in order to film my Zodiac bouncing in dramatic fashion up through the spume. As I revved the engine with a satisfying roar, the other boat struggling to keep level, the two of them hit a massive white-capped standing wave and the front of the Zodiac bounced skywards. The force of the impact threw Johnny Young up and out of his seat at the side of the boat, and as the RIB dropped again he came down sideways back on to the inflatable side of the boat, for all the world as if on to a bouncy castle. Then he bounced straight out of the boat and into the water, legs cartwheeling like a doll in a dishwasher. Purely out of survival instinct he grabbed on to the rope at the side of the Zodiac and was hauled along through the swell, bobbing underwater as he hung on for dear life. If he'd let go and been dragged back into the outboard engine with its roaring propeller ... well, it doesn't bear thinking about. We cut our engines and heaved him back on board. Though we laughed about it later when back at camp, it was a near-miss that none of us wanted to think too much about.

A week or so after our dreamy jaunt downriver we decided to go back downstream to the falls for a longer stint, in the hope of finding more wildlife in that pristine wonderland. Having returned safe and well, and wide-eyed with tales of the magical land we'd been exploring, suddenly everyone seemed to have forgotten their reluctance to make the journey and wanted to be on this next trip. Secretly I was extremely pleased with myself, not to say smug, to have been proved right about how benign the river is and how overcautious everybody had been.

The night before we left it started to rain. It began hard and got harder, and within an hour or so was truly biblical. Raindrops bounced angrily off the tarpaulins as if they were trying to punch through the plastic. As we all sat round eating by headtorch light, we had to shout to be heard above the constant roar. It was all quite cosy, though, beneath the camp shelters, and we all rather enjoyed the night off, knowing that not even Steve Greenwood would send us out to film in this.

Those of us who decided to get tucked up for an early night didn't last long in our hammocks, though. The river level increased by several metres in a matter of hours, and suddenly we were leaping up, roused by the chaos all around. A two-hundred-litre drum of fuel was swept away like an empty Coke can by the torrent, and forty-metre trees screamed past us as if the gods were playing pooh-sticks. Terrified tree kangaroos asleep in the foliage probably woke to find themselves being borne out towards the Pacific at the speed of a freight train. Our helipad completely disappeared, and for a week afterwards no one could take solace on the beach 'cos it wasn't there. Like a volcanic eruption, a huge avalanche, a hurricane, the river had become a destroyer of worlds ... it was one of those elemental moments when nature lets you know how insignificant you really are.

Weather events like this may be incredibly important in the seeding of species in island ecosystems around the world. Far-fetched as it may seem, huge trees really can be borne right out to sea, and then float for many days, weeks or months before they hit land. Some creatures such as reptiles can go into a kind of stasis, existing totally without food in the branches of the raft that such a tree becomes for them. Birds may use them as stopping-off rest points, fish hide beneath them as they are swept across the oceans, until they finally wash ashore in a new land.

Over the course of time, sufficiently large numbers of such Noah's Arks will find a destination, along with their living cargo, so that populations of breeding animals can set up on previously barren islands. Anyone who doubts this possibility should visit Krakatau off the west coast of Java in Indonesia. After the most famous volcanic eruption in history, Krakatau was blown off the face of the planet, and the land that remained was totally sterilised of all life. Go there today, and the thriving forests are filled with geckos and other lizards, snakes, birds, rats, bats and insects. Some seeds were doubtless dropped there in the faeces of airborne dispersers, and animals such as saltwater crocodiles and monitor lizards may have swum there. However, others are true founder populations, swept across from Sumatra, Java or even further afield on wave-borne flotsam.

*

While the river trip and the basecamp were essential elements of our series, offering extraordinary scientific discoveries, our main goal was to journey into the volcano that would give our series its name. Our basecamp was at its foot, and if you journeyed upriver for three or four hours you could see Bosavi's mighty bulk in the distance.

Our planned siege of the volcano and whatever was hiding inside it began in earnest about a week before we were due to set off. It was decided that we would be a three-man team. Myself, Tim Fogg and our cameraman 'Lucky' Luke – though after he'd been selected for the job of filming our full-on descent into God knows what, ultimately having to film the whole thing in a tumultuous downpour, his nickname would begin to look more and more ironic.

I was having a few problems of my own. Trying to assemble the fancy high-tech multi-fuel cooking stoves I'd brought from the UK, it soon became obvious that although they might claim to burn 'any liquid fuel commercially available in the world, including perfume and whisky', they certainly wouldn't burn our boat fuel. Every time I tried to light one it would send a dirty yellow flame six feet in the air, then splutter and die. After a day of very embarrassing tinkering, with all the crew, the locals and even the women who were there to do our laundry having a good chuckle at my expense, I quietly gave up, and stole Gordon's mini-stove that he used for brewing his morning coffee.

I've never been a big fan of technology on expeditions. The relative sanctuary of basecamp is practically bristling with satellite phones, GPS, satellite Internet hook-ups and endless modern jungle gadgets. Many of the guys will sit down of an evening, watch a movie on their laptop, try and usually fail to send emails, even play games on their iPhones. If it were possible, the two Johnnys would transport a whole film studio, viewing suite and electronic games room into the middle of the jungle, but it makes me want to scream.

The first and most rational reason is that I think on expeditions you need to streamline your kit, keep it simple and be self-sufficient. The problem with technology is that if you rely on it and it breaks down, you're stuffed. Use a map, a compass and your sense of direction correctly, and they'll never let you down, but if you rely on your GPS and you can't get a signal, you are irrevocably lost. There has been nothing more frustrating than the hours and hours I've spent getting up on to high ridgelines holding satellite phones up in the air, trying to get a single bar of signal so as to make a phone call, only to have it drop out after the first sentence. Technology is a fake comfort-blanket, it doesn't always work when you really need it, and if you'd not had a sat phone in the first place you'd never miss having to call home.

The second and slightly less rational reason I hate technology on expeditions is that it destroys the illusion. I want to believe I'm an explorer from a bygone age, miles from anywhere, cutting a fresh swathe across the rugged wilderness. To my mind that's half the experience of it. I want to be totally independent, relying on my wits and experience to keep myself alive. You can't even pretend that to yourself if your travelling companions have a high-tech beacon that can summon a platoon of paras to come abseiling in out of helicopters to your rescue within twenty-four hours.

There's also no doubt that this barrage of cargo alienates us from our local comrades. How can you haggle over prices with a local who's just seen you using a laptop that's worth more than he'll ever own? How can you establish true camaraderie with the local naturalists when they must see us as hopelessly profligate, not to mention helpless without our gadgetry? The indigenous contingent are the main reason basecamp works at all. For all the expertise and cerebral computing power of the science team, it is the quiet and tireless experience of our Papuan hosts that really brings in results.

A case in point would be our Papuan ornithologist Kwiwan, who is seemingly omniscient on the fledged fauna of the forests. Kwiwan is instantly likeable, with an easy smile and infectious laugh, but I wouldn't want to get on the wrong side of him: he has the naturally massive muscular shoulders, ludicrously low body fat and narrow waist of a true Melanesian warrior. He looks as though he could bench-press a Volkswagen. Kwiwan attempted to educate us all on the highland handshake. You shake as normal, but as you retrieve your hand you catch the fingers of your counterpart, making a loud click. None of the Western contingent can quite master it, and practically bring on arthritis just trying to learn. We compromise by shaking hands, letting go, then clicking our own fingers and hoping no one will notice.

While I can wander about in the forest for hours and come across nothing more than mosquitoes and termites, Kwiwan seems to always return with something spectacular. Luckily, he has spent time working for foreign companies and speaks excellent English. With many of the others it's not so easy, and we have to rely on our limited pidgin.

The lingua franca in Papua New Guinea is Tok Pisin, what the colonials called Pidgin English. The language made its way into New Guinea in the early 1900s, as workers returned from plantations in Samoa bringing a language that was a degradation of English. In a country with an estimated 841 languages, it took on the same function as Indonesian has in Irian Jaya, becoming a popular second language for many, one that can be used across boundaries of space and generation. Pidgin has become highly developed here, and if you try to communicate just with simplified English nobody's going to understand a word you say. Indeed, this way of attempting to communicate is seen as highly condescending, going under the name 'Tok Masta', the language of the white man.

Tok Pisin can be a wonderfully descriptive language. For example, 'I'm thirsty' translates to *Nek bilong mi i drai*, literally 'Neck belong me he dry'; a telescope is *bamboo bilong look look*, and *kalabus bilong susu* or 'gourd to keep your milk in' is pidgin for bra. You can get a good idea of the priorities of Papuan life from the fact that in my phrasebook the sentence *Nogut pik bagarup imp les matmat* comes many pages before 'Hello' and 'Thank you'. It means 'Don't let the pigs bugger up the cemetery.' The translations of Western concepts that are quite alien to Melanesians can be rather convoluted. For instance, a journalist is *man i raitim ol stori i go long niuspepa*, a lunar eclipse is *kerosin lamp bilong Jesus gone baggerap*, and a piano is *bigpela bockiss you bang im white teeth he sing, you bang im black teeth he cry*.

Returning late morning from an early foraging trip, when I walked into camp I became aware of unrest in the comms area of the kit tent. Much of the team were standing around the communications station looking thoroughly frightened. Steve, in the centre, was ashen-faced, holding the radio to his ear and sporadically shouting into the static: 'What do you mean they've hacked him up? How have they hacked him up? Is he alive?'

Static coursed back through the radio. 'Chocol! *Chocol!* What's happening there? How is Maxi? Is he alive?'

'What's going on?' I asked tentatively during a long pause.

'This is not good,' Steve responded. 'It sounds like the guys at Sienna Falls are involved in all-out war.'

Sienna Falls is one of the small villages on the slopes of Mount Bosavi; we were relying on them for permission to enter the volcano, as well as for their guiding and manpower.

'War? What do you mean, war?'

'Chocol's just been on the radio saying another village just burst in with machetes and axes. I can't tell if anyone's been killed yet, but it sounds like total chaos.'

Chocol was probably a little younger than me, but certainly destined to be the Big Man of Sienna Falls. It was clear he had more influence over the village than any of the actual elders. He had a powerful personality, an abrupt manner and perhaps an inflated sense of what he was owed. I had quite taken to him on earlier wanders, but Steve had picked him as a real troublemaker and didn't trust him one bit.

'Is there any idea why this other village have raided them, or did they just barge in all guns blazing?' This wasn't impossible, but it was more likely that there was a reason of sorts.

'Well it's very difficult to get a clear message.' Steve indicated the old-fashioned radio set. 'But it seems a little girl in Sienna Falls died of some mysterious illness. The people from the other village made up their minds that she had died 'cos the headman of Sienna Falls had put a curse on her. This other village are accusing him of being a sorcerer – dark magic and stuff.'

'So this dead girl wasn't from *their* village? Why the hell are they even involved?'

'I'm not sure why they're interested, maybe the kid's a relation of some kind.'

'That sounds pretty damn tenuous!'

'Yeah, it sounds like an excuse to cause trouble to me – anyway the funeral was today, and the men from the neighbouring village just barged in swinging machetes and axes.'

'Is there any word on Maxi?' I asked. Maxi was not only someone I'd got quite fond of, but also was supposed to be leading us into the crater.

'Well, it's all pretty fuzzy, but it seems Maxi got hacked in the arm. Chocol says the Sienna Falls men are all *belhat* and about to go over there and raise hell.'

This was not good at all. *Belhat* means 'belly hot', infused with fury, a state where murder is not only sanctioned but almost expected. As we listened on the radio we could hear angry yelling in the background, a woman's high-pitched scream, what sounded like the screeching of a person who had either received a mortal wound or seen one delivered to a loved one.

The screaming and wailing went on for some minutes, with no reply to our repeated calls on the radio. Then Chocol came back on.

'Maxi has gone!' he yelled.

We all looked at each other aghast. Not Maxi. This couldn't be happening.

'Stop, Chocol,' said Stevie Greenwood. 'Are you telling me Maxi is dead?'

Silence from the other end.

'Chocol, do you read me? I repeat, is Maxi dead?'

'No, Maxi gone,' Chocol replied. 'Maxi gone to get machete and go and kill 'em. This is big baggerup, Mr Steve. O-o-oh Maxi! You must send heli. You send heli, Mr Steve.'

In the background we could again hear the angry yelling, and what sounded like a scuffle.

Jane the medic and I instantly began mobilising to get ourselves to Sienna Falls and do what we could, maybe get some of the wounded out and even try to quell things with a little reason. Steve and Jonny Keeling wouldn't hear of it.

'We can take care of ourselves, Mr G. I can't just leave Maxi and Chocol to get hacked to pieces twenty minutes' helicopter flight away!' I pleaded.

Jane chipped in: 'Quite often you find that a female medic flying into a situation like this is just what it needs to make people see reason and quieten down.'

'And what if they don't?' Steve responded. 'What if you turn up, and this other village thinks you're siding with Sienna Falls, you'll be the first in the firing line.'

He had a very good point, but I was charged up to drop in there like Bruce Willis and save the day.

'What exactly are you going to do, Steve? Fly in there like something out of *Apocalypse Now* and hoist Maxi out? There's a whole village that'll want to be evacuated! The second you intervene, you're no longer an outsider, you're a target.'

'Imagine how we'll feel if they get killed, and we've sat here and done nothing?' I argue.

'Imagine how you'll all feel in a week's time when you're down in the crater on your own and there's a homicidal tribe out there that wants you dead. We're five days' walk away from them here and they're never going to come all this way to get retribution. Once you're inside Bosavi you'll be no more than a day away. You'd be totally vulnerable.'

This I hadn't thought of at all. The notion of machete-wielding Papuans appearing from the night forests when we were exposed and unprotected was chilling.

'We're not even going to get into the crater without Maxi and Chocol.' I reminded Steve: 'Maxi's a good man, we can't just sit here while he gets hacked up.'

'I'm sorry, but if Maxi's gone out with a machete to try and kill somebody, he is *not* a good man.' Steve is firm, reasoning, resolute. 'Maxi made his own bed the minute he picked up his machete. It's out of our hands now. I will not send you and Jane into a situation we know nothing about.'

'I guess one thing we could do would be to call Oil Search, have them send in their police,' Jonny Keeling muses.

'I guess we could *threaten* to send them in then,' Johnny responds. 'Maybe that'll scare them into calming down.'

Chocol came back on the radio: 'Mr Steve, you send heli, and bring

Maxi, very danger, suppose he kill some man? O-o-o-o-oh Maxi! You come *now* Mr Steve.'

As I mentioned, Steve had not been too fond of Chocol for some time now, believing him to be a bit of a troublemaker. Steve also doesn't like being ordered around, and has been running these expeditions long enough to swallow petty panics with ease while not taking any shit from anyone.

When he replied, it was stern and firm: 'No, Chocol. We will not send a helicopter. Maxi got himself into this when he decided to attack someone with a machete. This is Sienna Falls business, not ours.'

Chocol's responses were getting angrier – he started insulting us and our intentions. But in the background everything had gone quiet. Steve's categorical refusal seemed harsh to me at the time, but in retrospect he was quite right. My gung-ho attitude, riding in to the rescue like Roy Rogers, might have saved the day, but much more likely would have just created a whole host of bigger problems for everyone.

The conversation ended in a buzz of static, and everyone stood around tense and worried. Splinter groups formed in the kitchen and kit rooms, talking about what they thought should be done. The New Guinea old hands swapped horror stories about things that had happened when they'd become entrenched in tribal conflicts in the past. Jane and I started packing an emergency trauma kit, and she started refreshing my memory on trauma treatment and a few crude field operations. I had practised putting drip lines into my own arm, but I'd never had to do one for real. (The usual protocol is to take a partner and practise on each other, but on our last big expedition I got the medic as partner, and he decided to show off by using the heftiest cannula needle on the market – it was as thick as a pencil and looked like it had been designed for putting lines into shire horses. He pushed and shoved to break the skin, then abruptly stuck it straight through my vein and out the other side. A golf ball-sized lump instantly sprang up on my arm, blood pissed everywhere and my cameraman passed out cold! After that experience I was far happier, when necessary, just putting lines into my own veins; it's a good skill to have, as apparently whacking one in and venting a few bags of fluid into your veins is supposed to be the world's greatest hangover cure!

There was no doubting that Jane, a military-trained paramedic, had become totally focused since the radio call. I had the strong sense that she was actually excited by the scent of battle and danger, and the opportunity to do her job with real injuries rather than sitting around camp tending blisters, chafing and heat rash. I'm afraid I was infected by the same excitement.

After the initial spurt of adrenalin, the day dragged sluggish and sweaty, everyone on edge, working, but with their minds not really on the job. With nothing to do and too much zing in my blood, I spent a couple of hours exercising. To the total bemusement of everyone else on the team (particularly George, who teases me remorselessly about it), I always set up a little jungle gym near our basecamps. Not much, just a low branch for chin-ups, a sleeping mat for sit-ups and so on, an army duffel bag hanging from a tree stuffed with palm leaves as a punchbag, and a skipping rope.

When I returned, it seemed the drama at Sienna Falls was petering out. Maxi had got back there without having beheaded or eviscerated any of his neighbours, Chocol was sounding grumpy and disgruntled but no longer ordering Steve to take action, and Jane was packing and repacking her trauma kit for the tenth time, but with waning enthusiasm. In the space of just a couple of hours we had gone from the expectation of a massacre to general malaise and apathy. Such is the volatility and spasmodic ferocity of the Melanesian temper.

After midnight, Johnny, Nick and myself had just returned from a productive night walk and were sat in the kit room on the hotboxes that keep the cameras alive sipping Irish whiskey, talking bobbins and winding down ready for our hammocks. Just then, a burst of static on the radio: 'Stevie B, Stevie B, this is Gordon come back.'

'Hi Gordie, it's Steve here, we're just finishing off the last of your whiskey.'

'*What*! You buggers! Look Steve, I'm out here filming, and this big white snake is hunting right alongside me, we think it might be a small-eyed snake.'

I paused over the handset, trying not to look at Nick and Johnny, who had their faces in their hands and were groaning loudly. We were all washed and wearing our dries – getting back into our daytime wets and traipsing off into the darkness after a very long day's filming was not on our wish list. However, the small-eyed snake is one of the most lethal members of the cobra family found in New Guinea, and I'd been charged by the internationally respected herpetologist Dr David Warrell to collect a venom sample from a small-eyed snake in the area. None had ever been milked here, and the venom could be vital to his studies.

'What's your approximate location, Gordie,' I asked, fingers crossed that he'd tell me he was just outside camp or at the riverside.

'We're on the C trail spot near the remote cameras by the cassowary pond,' replied Gordon, to more groaning from Nick and Johnny. 'Guess it's about forty minutes from camp.'

Grumble they might, but the crews all understand that no matter what hour of the day or night, if there's an animal to be filmed they have to be up and out, so within five minutes everyone was ready and we were off. Needless to say, we pretty much sprinted all the way out into the forest, certain that we'd turn up and the snake would be long gone. If that did happen, at least we'd be able to turn around and be back home before Tony the cook discovered the last of our whiskey. But when we reached Gordon out in the forest, he pointed straight down to a nearby tree, and there it was.

On the face of it, the snake was without doubt a small-eye. Not closely related to the Australian small-eyed snake, which is dark and unremarkably coloured, the New Guinea small is a visually extraordinary creature. Natives describe it as a 'long white snake that seems to shine in the dark'. The snake in front of me fitted the bill perfectly, with its shiny white coloration standing out vibrantly against the brown leaves – almost as if someone had thrown an ivory walking-cane down on the forest floor. It also had the distinctive golden streaks and bands, and its beady black eyes were indeed small.

This stunning serpent was moving steadily around the base of the tree, burrowing its head through the leaf litter, clearly hunting, constantly on the move in its search for a meal. However, there was something about it that bothered me. Perhaps the eyes weren't quite as tiny and recessed as I'd seen in pictures – I'm not sure. But there's one way to tell for sure. The elapids, the members of the cobra family, all have the pre-ocular scale (the one in front of the eye) and the nasal scales directly contacting each other, whereas in other groups these scales are spaced out with another scale in between. Taking account of the dentition – number of teeth, alignment and so on – is also essential to figuring out what kind of snake it is.

To assess both those things, you need to catch it. As I went to pin the snake down with my snake stick, however, it all kicked off. The soft dead leaves meant there was no purchase to pin it down, and he just wriggled free of my stick and thrashed off at a great rate of knots, then striking viciously back up at me and lunging between my legs. The other guys gasped in horror – to be bitten by a small-eyed snake out here might not be *certain* death, but it's no exaggeration to say it would be *probable* death. Its venom attacks both the muscles and the nervous system, within minutes you are unable to even stand, and your breathing and heart start to shut down. The venom contains anticoagulants and haemolytic factors that lead to massive internal bleeding – the victim passes almost black urine as his muscle cells are dissolved and then eliminated as waste products. Even if you survive (and most of the case study patients I've seen died within the day), the effects would last in your body for many years.

As the snake kicked away at a pace, all my senses were telling me to give up and let it go, it wasn't that important. But then I looked up at Gordon, Jonny, Nick and Johnny, people I respected and wanted to respect me, looking so expectant, and decided I couldn't be so publicly defeated.

I went in for another try. Again, he struck at me repeatedly, writhing round and round. I got a hold of him just above the tail, but had to let go as he went for my hand. This was turning into the most nerve-racking and least graceful snake catch I'd ever done. Finally, he tried to wriggle over a solid root, giving me the purchase I needed to pin his head down with the snake stick and clasp it between my fingers, taking great care not to lose control and get jabbed by a fang. With a venom this toxic, even the merest of scratches could inject a lethal dose.

I was shaking with adrenalin as Gordon and I sat down with the snake to identify it for the cameras. The scales behind the nose told the first bit of the tale. Then I opened its mouth and found no long slender cobra fangs waiting to inject me. My heart rate slowed, as

I checked and rechecked my diagnosis. But there was no doubt about it. Our evil beast was not a small-eyed snake at all. It was actually a rather spectacular ivory-coloured morph of the common ground snakes that we'd been finding around camp every night. All the way back to camp I had Johnny's and Nick's eyes burning holes in the back of my head, knowing they'd only get a couple of hours' sleep because of my ludicrous obsession with nasty things that bite and sting.

It's a shame we only found one ground snake with this startling coloration, though, because any more and I'd have begun to think this was highly significant rather than just a freak colour variant. The ground snakes around camp (and we'd caught many) were all nondescript brownish, and perfectly camouflaged in the leaves they were usually seen wriggling through. No doubt their main protection against predators such as monitor lizards, cassowaries and eagles would be crypsis – simply the ability not to be seen. However, this shiny white snake stood out in the undergrowth like a drag queen in a Yorkshire working-men's club. So why hadn't it been eaten? Well, because it looked exactly like one of the most venomous snakes in the world ... could it be that other predators had shied away as I so nearly had?

In biology, mimicry is one of the most fascinating of subjects, and it's particularly well known in snakes. The classic examples are found in South and Central America, where highly venomous coral snakes sport brightly coloured bands, referred to as aposematic warning coloration. Other animals learn to leave them well alone, but certain harmless snakes such as king snakes have over time evolved the same coloration as the venomous ones, and hence are also left well alone. The mimics have the dual advantage of being safe from predators while not having to pay the biological cost of creating venom, which is actually quite demanding of an animal's energy.

Examples of such mimicry abound throughout the natural world. It can be used aggressively to con potential prey as well as deter predators, and is sometimes even to the advantage of both parties. The classic case of this is the foul-tasting and brightly coloured monarch butterfly. For many years the almost identically coloured viceroy butterfly was thought to be getting a free ride off the monarch's warning colours – until it was realised that actually the viceroy tastes even worse than the monarch! This benefits both species, as more and more predators will be trying them, spitting them out and learning to leave bright-coloured things alone.

It would be too much to suggest, from one specimen of a bizarrely coloured snake, that we were seeing mimicry evolving in ground snakes there in New Guinea; after all, evolution works on populations of animals over time, not on individuals. But it was a tantalising possibility, and kept my mind occupied as we yomped home and to a so-welcome bed.

With just a few days left to go in basecamp, myself, Tim Fogg and Lucky Luke our cameraman began our preparations for the big push into the crater of Bosavi Mountain. Chocol and Maxi had returned to basecamp as if nothing at all had happened. Maxi was again his cheerful, giggly self, but Chocol, on the other hand, was beginning to prove Steve Greenwood right, and turning out to be thoroughly arrogant, self-opinionated and difficult. He barked orders at everyone, told us how things were going to be, and made it clear that he believed Sienna Falls was not being paid nearly enough for allowing us into Bosavi. There was going to be real trouble when it came to the final negotiations for our departure from New Guinea.

For now, though, we needed Chocol onside, and we also needed his help, so we ignored his pontificating and got on with sorting things out for the trip to the volcano. It was difficult to know where to start, as although several Australian prospectors and district officers had skirted the edge of it several decades back, and one naturalists' expedition had ventured on to the outer slopes, we were going to be the first outsiders to enter the lost land inside the crater. Chocol told us that despite Sienna Falls having ownership of one side of the mountain, nobody from his tribe had been there in years.

'It is too hard walking there,' he said. 'O-o-o-oh very steep, very tricksy. Take many, many days walk down into Bosavi Mr Istiv. But the animals ... tsk-tsk-tsk.' He shook his head in disbelief, then went wide-eyed with excitement: 'The animals is many many, Mr Istiv, and they are not frightened. When Sienna Falls people have big singsing, we wait and go into Bosavi. Very good tree kangaroo, very good echidna, very good eating meat.'

'And have you been into Bosavi, Chocol?' I asked him.

'Many many times,' he said confidently. 'I know Bosavi better than the other men. They say they know Bosavi, but they are liars.'

'And when did you last go there?'

He pondered. 'Hmmm. It was some times ago. With my father I walk right through im Bosavi, maybe take four days, five days.'

'Some times ago? What – years, months, weeks?'

'Some time Mr Istiv. When I was small boy.'

Ah. Quite some time ago then.

'And when do you think the last time anyone went into Bosavi was, Chocol?'

'Two year. Maybe more. There is no path there now – very tricksy . . . tsk-tsk-tsk.'

As Chocol didn't really have any information as to how long it would take and where exactly we should go, my plan was to fly the helicopter right into the bowels of Bosavi, having first spotted from the air something that looked like a potential landing site. If we found something that we could risk landing on, then perfect, but that seemed highly unlikely. More probably, we'd look for a potential site from the air, mark it on the GPS, then land on the outside of the volcano, trek up to the rim, then down into its heart to try and find the landing site. It wasn't quite throwing darts blindfold, but not far from it.

As we would have to carry on our backs everything we needed, it was essential to keep the weight down. We each carried a simple hammock with a tarpaulin to keep off the rain, plus a sleeping bag and a change of clothes. I took Luke's personal stuff so that he could carry the camera equipment. Then there was a stove, a pan, some freeze-dried food and a medical kit. Maxi took an axe and we all had machetes for clearing the way and for cutting the helipad. Not all of the kit-rationalising went to plan, though. I went to break the handle off my toothbrush to get rid of a little bit of unneeded plastic, but unfortunately my thick sausage fingers did a less than spectacular job. It split right in the middle of the bristles, leaving me with nothing at all. I had to clean my teeth with my finger for the rest of the trip.

Waiting on the beach for the helicopter to come and whisk us away was particularly sweet. The flies had been at their worst over the preceding few days, turning every second in basecamp into a buzzing Hades. Good animal encounters had pretty much dried up and everyone was looking haggard and harried, but Tim, Luke and I were bound for a magical wonderland.

It is one of the biggest frustrations that our journeys into the heart of New Guinea, which had the aim of drumming up the political will to save these besieged territories, would not have been possible without helicopters. Unfortunately the only ones available to us were owned by the very people responsible for the rape and pillage of New Guinea's forests. We'd drawn the line at using logging companies for support, but had unfortunately had no choice but to use Oil Search, the energy company whose major stakeholder, as mentioned earlier, is Exxon Mobil. Their main base is at a place called Moro in the southern highlands on the banks of Lake Kutubu, and we needed to pass through Moro pretty much every time we went anywhere. It was a very strange place, out of kilter with everything else in the highlands. Here, gargantuan double-bladed Sikorsky helicopters ferried in underslung containers the size of bungalows, and pot-bellied geologists from Wyoming in Ray-Bans and cowboy boots drove around in shiny white Toyota Landcruisers.

One of Moro's largest projects is the opening up of a major gas field under Lake Kutubu, an internationally significant wetlands habitat. The liquid natural gas is destined to be pipelined back to Port Moresby, and according to some sources is worth as much 11 billion US dollars. In 2007, while the Moro branch of Oil Search was drilling near the lake, locals reported that dead fish were floating to the surface and that people who ate fish from the lake or drank its water got sick. After falling sick just hours after eating perch from the lake, one local girl died two days later. Despite the conviction of local people that the pollution was caused by run off from the drilling there, no proven link has been established. All the same, we felt a bit ashamed to be in that place at all. Everything it stood for reeked of multinationals exploiting the Papuans and their lands.

While in Moro we stayed in prefabricated units, sealed plastic rooms with no windows, and beds covered with unspeakable stains. Alcohol was banned, as was straying outside the high razor-wire-topped fences.

Originally Moro had been locally owned land – Oil Search had bought it from the tribal Big Man. The tribe that had owned all this land had once lived here hunting the forests and fishing the lake. They now lived in an unbearably sad little shanty-town beyond the wire barrier, and stumbled about dead-eyed, in shabby clothing. We were warned that to venture beyond the fencing to one of their little kiosks or to the village pub would result in assault or even worse; violent gang rapes by *rascols* are a daily occurrence in the urban centres of New Guinea. Most of the men and women who came near the wire, staring through the fences at us, were clearly drunk all day. I'm not sure who felt more like the animals in the zoo, them or us.

Nothing about Moro seemed real to me. The canteen served school-dinner-style Western meals, and you could make yourself a Mister Whippy ice-cream from the dispenser as you left. Everyone, from the Wyoming geologists to the frizzy-haired Papuan workers, studiously curled the ice-cream out with intense seriousness, trying to get the maximum possible into each cone. It was pure comedy to see these hulking great workers in their greasy dungarees balancing foot-high Mr Whippys as they left, white trails dripping from their curly beards.

The people of Oil Search have to battle constantly with the fact that – in sharp contrast to Irian Jaya – land in Papua New Guinea is still mostly in the hands of the original owners, and the Papuan government stipulate that foreign companies have to approach these landowners if they want to use their land. This means that Oil Search are always having to impress the locals in order to sweeten deals and make them give up their booty more easily. Every so often a shining new helicopter will drop on to the landing strip, and a chieftain in loincloth and headdress of cockatoo feathers will emerge, blinking and incredulous, but nonetheless impressed and ripe for picking. On one occasion, Johnny Young and Steve Greenwood were sat waiting at the airstrip, and a helicopter landed not only with a local Big Man on board, but also with the cassowary he'd brought along as a gift for the men he'd be bargaining with. At this point, the cassowary broke its primitive shackles and sprinted off down the runway, with a furious chieftain and three Oil Search lackeys in tow, trying their best to rugby-tackle a terrified six-foot bird that has been known to kick its way through a riot shield!

But the bizarre, almost comical incongruities at the Oil Search facility were soon to be exchanged for a scene of horror.

It was the afternoon of an interminable day waiting for the all-clear to head off into the mountains, and I was sat on the porch of our block with a few of the crew. We had to sit out here because we'd washed our hammocks and boots out and hung them outside to dry. The warden had told us that anything we left unattended for a few minutes would be stolen by the Oil Search workers, so we sat in a bored vigil, not quite as bad as waiting for paint to dry, but almost.

Suddenly, the shanty-town on the other side of the fence came alive with violent screaming and yelling. A Papuan woman with shaved head, wearing just a dirty white bra and skirt, was dragged out of one of the huts by two men. As we watched, we were horrified to see the men take turns to punch her full in the face, and as she doubled up they punched her in the head, then laid into her with mighty kicks.

One of the guys was yelling in a kind of pantomime fury, shaking his arms about his head, bunching his fists at her and rolling his eyes. Over the top and unreal, it was the bizarre display of two wrathful males and their terrified subject. There was, however, nothing fake about the punches and kicks they were throwing. The woman was screaming in a guttural and deeply unsettling way. A crowd was beginning to gather around them. Fighting and kicking to get away, she fell on her back on to the ground. One of the men grabbed her leg and dragged her through the dirt. Her tattered skirt rode up, showing her nakedness, and her fingernails scraped along the ground. She was trying desperately to escape, all the time screaming for mercy. Then two hard kicks from one of the men landed with a sickening thud on her face, clearly audible where we sat. She lay on the ground shouting and weeping, holding her face in her hands.

'Holy shit!' said one of the guys in horror. 'They're going to kill her.'

We stood for a moment in silence. 'What the hell are we going to do?' We were on the other side of the fence, and unable to intervene.

'I'm going to get some help,' I said, pulling on my trainers and heading off in the direction of the office.

Just before turning to go, I saw something that will always stay with me. One of the men picked up a hunk of concrete the size of a football and threw it down on to her head. It connected with a hideous finality, and she went very silent and very still. With the adrenalin rising, I ran round to the warden's office, pushed my way inside and started shouting for help. I'd just seen a woman being killed right in front of me and been able to do nothing at all about it!

Some minutes later, the warden came out to the hatch that separated management from the workers. He was a Papuan, tall, very dark, with brown betel-stained teeth and an arrogant, cruel demeanour.

'You have to help, please, they're killing a woman outside, on the other side of the fence!'

He looked at me, excavating his teeth with a toothpick.

'So?' – then almost a laugh – 'What would you want me to do?'

'I don't know – stop it, get out there and save her, they're *killing* her!'

This time he laughed unrestrainedly. 'Suppose they *are* killing her? She probably deserved it. This is not my problem, they sort out their problems their way.' Another hearty laugh that chilled me even more than the sight of that rock staving the woman's skull in.

'Please, there must be something you can do, send the police out there, do *something*!'

His grin now had a tinge of annoyance about it.

'Don't you tell *me*! You white people, you come here, you understand nothing. This is Papua. She suppose sleep with someone who is not her husband. You ask me do something, I tell you *No*, this is not to do with me. Do not tell me something you do not understand.'

At this point I could feel the bile rising, helplessness, impotence, anger.

'Right, well, I'm going to go out there and stop them if you won't. I'm not standing here and letting a woman get stoned to death. No way!'

Now he turned serious. 'It is not allowed to go outside the gates. You do not understand anything. If you go out there, it is *you* they will kill.

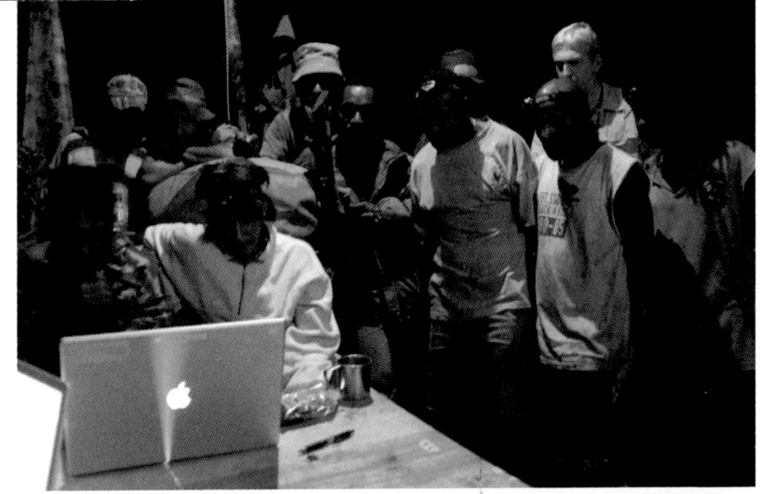

Above The local guys join Jane the medic (and Steve Greenwood peeking over from behind) in watching *Lord of the Rings*.

Right Nick the soundie and me enjoying a chuckle whilst dreaming about beer and pizza.

Below Steve Greenwood dwarved everyone in Fogomayu village – some of the adults barely reached his midriff!

Top Though many of the adults in New Guinea can be mistrusting and keep their distance, kids are much less reticent. Often the best way to break the ice is by playing silly games.

Below At a Sing Sing inside Fogomayu's ceremonial house.

Top The glory of Mount Bosavi from the air. It's only from this height that you can begin to appreciate the dramatic caldera.

Below Mount Bosavi as seen from a distance.

FACING PAGE

Main picture Having finally made it to our potential camp site in the depths of Bosavi, Chocol and I set to assembling the beginnings of a camp.

Inset Tim Fogg jummaring up one of the vast trees inside basecamp.

THIS PAGE

Top Tim and me in the gully that ended in an impassable waterfall as we pushed on towards camp Bosavi.

Below Megabrain George McGavin and I share a cup of coffee on the rim of Bosavi.

The Bosavi giant woolly rat, the largest rat on the planet, a new species to science, and only occurring in this one isolated mountain.

From left to right, Chris Helgin, Robin Smith and Gorgeous Gordon Buchanan, looking chuffed with themselves after finding the giant rat.

Looking down at the green tarps of camp Bosavi, the ferns in the foreground made a rather lovely carpet.

Bosavi camp, eternally ankle-deep in mud, plagued by leeches and mosquitoes, but home for three weeks of extraordinary exploration.

Clockwise from top left 1. Sat at the front of the RIB, pretending to be in the spotter's position, but actually only here as I've been banned from driving the RIBs after nearly sinking one in a whirlpool. *2.* A thundering cataract in the land of the falls below basecamp. This stretch of river, with mammoth falls rampaging into the main river, is one of the most beautiful places I've ever been and few outsiders have ever seen it. *3.* As we travelled down into the land of the falls, great white thundering curtains of water tumbled down to the waterside with astounding regularity. *4.* Beginning my ascent of one of the cataracts in the land of the falls.

They are *belhat*, this is the Papuan way. When they are like this there is nothing you can do. You want to get killed for some cheating woman?'

'She's a human being! And she's being killed on *your* patch! I will go and talk to the big boss and tell him you did *nothing*. You think he will like that?'

'You go, do that. I am busy man.' But his attitude was different now, and he looked at me thoughtfully. 'But suppose I am warden, and I am policeman here. Suppose I get men go look see.'

He paused as if this was the end of the matter, then waved me away with the back of his hand. 'You go now.' He turned and went back into his office, and I sprinted back to the porch.

All was quiet. The guys told me that the prostrate woman had been dragged off into one of the huts. Dead or alive, they couldn't say for sure. Several of the Papuan Oil Search employees had gathered round the wire near us to gawp, laughing and joking as if they were at a football match. I went over to them to try and find out what was happening, and one of them explained the situation.

The tribal Big Man had been away for several months visiting other tribes for *tok tok* about lands Oil Search wanted to buy. While he'd been away his wife had been sleeping with another man. He'd returned and found out – and then gone *belhat*. Like the warden, they seemed to think it was all a joke, funny, inevitable. They laughed at what had happened to her as if telling jokes about something that was nothing to do with them. It was deeply, deeply chilling. They didn't strike me as being cruel men – indeed, they had smiley, friendly faces – but it was just as if the woman was nothing at all, as if her life was of less consequence than a fish or an insect.

We never saw any police from Oil Search head over into the village, and never heard what befell the woman and her lover. You'd have to say, though, that it didn't look good for them. Shortly after we returned to Britain, it was reported in the international news that within the space of a few days, at around the time we were in Moro, three women had been killed in our highland province – stoned, then dumped in their huts. The huts were set alight, and they were burnt alive. Perhaps most frighteningly, it seems our woman may not even have been amongst their number because those women had all been accused of sorcery and burnt as witches.

The Dark Island had again confirmed its propensity for extreme violence.

14

In the musty dark of a tambaran, *or spirit house, a man caked with red clay and wearing a massive, quivering headdress, chants an ancient story and dances to a hypnotic rhythm, skipping from foot to foot. Smoke from the smouldering hearth drifts around the clustered faces of the clan, eyes white in the gloom. Mothers nurse their babies at the breast and children sit in rapt attention, staring more at me, perhaps, than at the dancers. The man's fingertips, coated with a sticky resin, pluck a strangely resonant chord from his dragon-skin drum. With every step, cascades of nuts at his ankles and the bracelets on his wrists rattle like castanets. With four others he holds the floor in front of us, as we sit with eyes streaming in the ghostly, smoky firelight. This is singsing, the ritual of convening with the spirit world and with ancestors, to find favour for our journey and invoke the blessing of nature to guide our way and keep us safe. Any inauspicious signs, and we will be unable to raise a single man to help us, for they will know that taking any part in the mission would spell misfortune . . . even death.*

Our little team assembled upriver of the basecamp at the village of Fogomayu, nicknamed 'foggy' 'cos it was – foggy. The village had a population of only a few hundred but was spread out over a sizeable area, surprisingly far back from the river, with a grass airstrip surrounding vegetable gardens and a big thatched house they'd built specially to accommodate us. As we were bringing money and trade to the people, the headman of Foggy insisted on having a singsing, at which several pigs would be slaughtered and eaten and we would be party to a big tribal dance in the clan house.

As the day of the preparations wore on, people started arriving from the surrounding villages, all dressed in their traditional garb. The women were clad in grass skirts and headdresses, but just as the male birds of paradise totally outstrip their female counterparts in glamour, so the appearance of the men of Foggy and its environs was far more dramatic. They were bare-chested, displaying their impressive musculature, and had covered themselves in red mud. They wore headdresses that included a cap of cuscus fur, and feathers from cassowaries and birds of paradise.

In his early expeditions to New Guinea, David Attenborough met with several thousand dancers at a singsing in the central highlands. From the number of birds killed to clothe a single dancer, he reckoned that one singsing represented the deaths of forty thousand birds of paradise. Nowadays, several NGOs mount campaigns, sending representatives into villages armed with plastic bags and mothballs, encouraging the dancers to preserve their old costumes rather than kill more birds to make new ones. It's highly unlikely that this effort will succeed before at least some species of birds of paradise are extinct.

For New Guineans singsing is a serious business, and goes on for several days. They dance all through the night, the hypnotic rhythm and repetitive refrains seeming to send them into trances that more than one contemporary anthropologist has likened to the effects of modern dance music parties and raves.

As anyone who's followed the experiences of anthropologist-explorer Bruce Parry will know, many tribes around the world supplement their rituals with hallucinogens, but here in New Guinea the strongest drug available at a singsing is betel nut. As noted earlier, betel is the seed of the areca palm, which is mixed with areca leaf, maybe some tobacco leaf and some crushed lime. It's chewed in many Asian and Oceanic countries, and is responsible for pavements appearing to be blood-spattered and for many regular users bearing blackened teeth and gums. The alkaloids in the nut react with the lime, creating oodles of blood-red saliva, which is spat out. It gives a mild high, ups the heart rate and makes you feel hot from inside. It is also one of the most vilely bitter tastes imaginable, but is highly addictive. Many cultures in this part of the world assign symbolic meaning to the chewing of betel. The nut itself often symbolises the ovaries, the lime signifies sperm, and the resulting red spit is blood and the life that issues forth from the blending of the egg and the sperm.

The singsing was a curious affair. It took place in a particularly dark and musty spirit (or clan) house, packed with women and children, though most of the serious dancing was done by the men. The dances varied. Several muscular young men performed aggressive moves, leaping from one foot to the other while beating on a long, carved wooden drum. The skin of the drum is made from cured monitor lizard skin, and the players have a sticky resin on their fingertips so that as they strike the drum skin and then pull away, the fingers stick and create an unusual and eerie resonance. They burn the resin and heavily scented incense with flaming torches, the oily substances fizzling in puffs of fire that cast dramatic lights and shadows across the starkly made-up faces. Their chants are chillingly fierce.

Their writhing physiques shine with sweat, their heavy brows furrow, eyes glint ferociously. The whole display looks like it would send a troupe of haka-performing Maoris scampering away with their tails between their legs. One particular young warrior seems to be glowering in my direction, and I have to keep reminding myself that this is not a challenge but an offering, an invocation. Then the younger men step aside, and Chief Seguro from Sienna Falls village steps in. Seguro is a powerful, domineering character, probably in his late forties; the piercing in his septum is so pronounced that you can see clear through it. He dances alone, telling the legend of the creation of Sienna Falls.

His song and dance go on for much longer than the dances of the young men from Foggy, and after maybe twenty minutes the man performing before us is wearing a look like thunder. How dare this chief of a smaller village come into *his* singsing and take over? He seems to be asking. Abruptly the younger men interrupt Seguro's routine, and leap in with their own hopping drum-beat mantra. Seguro does not step aside, and the second the young men take some respite he steps into centre stage and begins again the same monotonous chant. He is painted red from head to toe in ochre and pig fat, and his headdress of cassowary feathers hangs over his head like a cock's crest, nodding up and down with every bounce. He closes his eyes in rapture as he sings his story for the white men.

Now the young men of Foggy are looking furious. They interrupt again, this time before Seguro has completed his verse, turning their backs on the older warrior, ushering him into a darker corner of the spirit house and drowning him out with their more combatative performance. The tension fizzes like static. I'm no expert on the finer points of singsing, but this looks as if it might turn ugly. It seems Steve Greenwood has picked up on the same vibe, for he chooses this moment to stand up with his hands held high, call proceedings to a momentary halt and through the *turnim tok* (translator) explain to the representatives of all the assembled villagers how much he values them and their goodwill. The chief of Foggy announces to the gathering that the film we will make will bring more foreigners to the area and usher in a new time of prosperity. Though we all look a little concerned at this, the tension has been broken. It would perhaps not be a good time to say: 'Actually, no, nobody is coming here to Foggy. Possibly not for many many years – it's certainly *not* going to turn into Benidorm'!

Though the singsing continued for some through the night and the next day, the members of the Bosavi expedition all slunk away at various points in the evening. There's no getting round the fact that if you don't understand the stories and are just sat there watching and not dancing yourself, these things can get a little wearing after four or five hours. Also, we had preparations to make, and knew that there would be days ahead when sleep would be a lessening and lusted-after commodity.

The hard work over the next few days came in an unexpected form for me. Fogomayu village was suffering from a malaria epidemic, and Jane the medic decided to take matters into her own hands, setting up a temporary clinic to treat the sick. Within minutes of word spreading that a healer was in town, a queue of twenty or thirty people formed, suffering from everything from cuts and scrapes to the aftereffects of advanced cerebral malaria. The children who had been through this terrible disease were drooling and could barely stand, their eyes rolling back to the whites, evidence of the severe brain damage we could do nothing about. Few people in the village spoke any pidgin, so we used one of the locals who did, to translate, and then Kwiwan or his friend Musee would translate into English.

There were scores of very young babies with their bewildered and frightened-looking mothers, screaming with fever; or, more worryingly, silent. Jane was in need of a helper, and I was desperate to put my remote-area medic courses to use, so volunteered. Jane was truly inspirational, showing me how to measure up tiny portions of quinine tablets to give to the infants and inject suffering adults in the backside with the drugs. The first few injections I gave were a little hesitant, and it was obvious from the patients' faces that my jab hurt more than the malaria – they would rather be stuck by the expert, thank you very much. I did manage to find my niche, though, when several old men turned up suffering from back and knee pain. Having spent years playing contact sports like judo and rugby, my joints are all pretty much knackered – I've had more physio than these guys had eaten hot cassowary. That, plus a rudimentary sports-massage course I'd done many years before, meant I got the job of sorting out the old men's aches and pains.

This involved me actually laying them down on a tarpaulin out in the sun, getting some cooking oil from the kitchen, and just basically giving them a good old-fashioned rub-down, while half the village looked on and rolled around in hysterics – to the misery of the elders. It was all pretty unpleasant, oiling the leathery-skinned and slack-muscled old guys, their genitals lolling out of their ancient shorts and loincloths; especially as washing here is not a particularly regular occurrence.

However, the results were remarkable. When one old guy came in with terrible back pain, it was immediately obvious that the crutches he was using for his bad knees were way too short and were bowing him at the waist, thus causing him to bend double and inevitably wrecking his lower back. All I had to do was cut some new crutches, give him a massage and half an aspirin (pretty hardcore pain relief for these people, who'd never had a modern analgesic), and he was so relieved that he almost wept with gratitude, and pumped my hand up and down. He might as well have thrown his crutches in the air and skipped with joy! I for my part was beside myself with gratitude to Jane, who had given me the opportunity, for a couple of days at least, to feel like I was actually doing something to help people. Despite all the pain and tears we witnessed, it was one of the most uplifting experiences I'd ever had.

On the afternoon of the second day of our clinic, things again turned sour. A short, plump Papuan with a well-tended moustache and a pot belly (obviously from being well fed rather than from the malnutrition that so many of the kids suffered from), wearing a big shiny watch and unusually fresh-looking Western slacks and polo shirt, turned up at our makeshift clinic and started making a scene. He was angry, shouting at the women waiting in the queue with their babies – obviously trying to break up the clinic. He stood right next to Jane as she attempted to work, and in the local dialect denounced everything she was doing.

The queue began to thin out. The man's intimidation tactics were clearly starting to work: all of a sudden the women wouldn't meet our gaze, everyone went silent. Although he wasn't addressing us directly, things eventually became impossible, so we got Musee to translate for us and find out what his problem was. I had kind of assumed that he was some modern equivalent of the witch doctor, who would be the most important of the village Big Men and a powerful ritual presence in the traditional Papuan village. In my romanticised vision, he mixed potions and chanted voodoo magic in a hut filled with spirit masks. Therefore our newfangled drugs and twenty-first-century witchcraft had offended his sensibilities, and we would probably be cursed or have our heads shrunk for us.

'Fogomayu is his village, it is his job to help these people, not yours. He tells you to stop, or he will be very angry,' the *turnim tok* told Musee, and then Musee told us.

'These babies have malaria, if we do not treat them they will die,' Jane calmly replied.

'He say he knows this, he treats these babies all the time with the quinine.'

Ah, so he does have medicines, not just chants and amulets.

'Where does he get his drugs from, then? And how does he know how to use them?'

'He is a doctor, Oil Search pay him to stay here, and to give the medicine when we are sick.'

'OK, and does he treat you?'

The *turnim tok* shifted from one foot to the other, but then obviously realised the Oil Search guy couldn't understand the translations – 'No, Missus Jane, he never give us nothing. He have a clinic down towards the river, but it is never open. He is a very bad man.'

'When was the last time he ran a clinic?'

Jane's question was responded to with a shake of the head, which clearly indicated if there had ever been a clinic, it was many months in the past.

'So this guy is paid a salary to be here giving these people medical care, and he's doing nothing? What a scumbag!' After just over a day of helping I felt like Florence Nightingale, and was ready to come to the aid of my flock.

'Tell him we're treating these people, and if he doesn't get stuffed he'll have me to deal with!'

Jane raised her hand to make clear this last outburst shouldn't be translated. Luckily, she'd dealt with much worse in the past, and knew the best ways of getting round such men.

'This is a very bad man, Missus Jane,' the *turnim tok* offered. 'He take the medicine Oil Search give him and sell them to other villages. Also, he hate the women and won't treat them. Last week one woman had a miscarriage, he would not give her any medicines and wouldn't call Oil Search to fly her out with the helicopter. She died. This man is very bad.'

It was obvious what had been going on from the faces of the few women who remained, standing there with eyes downcast and the screaming babies in their arms. They were desperate. They hated him, but could do nothing. He had the power of the oil men and their military police behind him, and they feared that worse than disease. Add to that the fact that the village headman was also in thrall to Oil Search, believing he was going to make big money out of selling his lands to them. The villagers' situation looked grim.

Jane's response was a canny one. She merely said through the interpreter: 'We're really sorry, we obviously don't understand about medicine in Papua. With your kind help, would you teach us?'

This totally threw our angry medic. He just stood there with his mouth opening and shutting.

'Tell him that there are many people here who have not been treated, and our friends at Oil Search will not like that. Tell him we'd like to see his clinic – that way we can tell our good friends at Oil Search what a hard job he's having here. Maybe we can get them to fly in on the helicopter tomorrow and see for themselves.'

This took the wind out of his sails. Jane had just stuck a pin in this over-inflated balloon, and he started making excuses, even proposing that we could continue with our surgery if it made us happy. But Jane insisted. Eventually, with a crowd of women and children following, we were led by the crestfallen medic, tail firmly between his legs, to the medical hut. It was the most modern construction in the whole village, with corrugated-iron roof and creosoted wooden walls, but the windows were boarded up and the doors locked.

At first he claimed to have lost the keys, but Jane was too tough to give up that easily and after a while he acquiesced, and the keys magically turned up. He let us inside. The place had been kitted up pretty well, with ancient posters on the walls instructing how to use condoms to defeat AIDS, and how to wash hands to beat disease in general. But there was a thick layer of dust over everything, and the consulting room was being used as a storeroom, stacked high with bits of boat motors and old roof thatch. The modern metal treatment table was buried under boxes, which looked as if they had been used to roost chickens. All in all the whole place looked as if it had never been used.

At Jane's insistence, the reluctant medic opened up the store cupboards to reveal clingfilm-wrapped stacks of the very drugs we'd been using to treat malaria in the village's babies, along with maybe twenty other drugs to treat giardiasis, dysentery and other infant-killers. These were packed in bulk-delivery boxes, all unopened. Some of them showed expiry dates several years old. It was a total travesty. Here in this cupboard was all the simple modern medicine it would have taken to relieve the pain we'd seen over the last couple of days, and the key belonged to this repulsive little tyrant. I glowered over him with clenched teeth while he stared at his feet, and as Jane told him how wonderful his facility was and how she would be ensuring that Oil Search found out exactly how their drugs were being used. It was clear to me that Oil Search can have had no idea. Even with the barrier of language, it must have been clear to him what was being demanded.

The next day when we were preparing to leave, the surgery was open to business, and not just to Fogomayu's male population. A long queue of women with their babies tailed out of the clinic and on to the airstrip. The rest of our crew flew out to Moro, taking with them a critically sick woman with her husband, referred to the Oil Search hospital by our charming medic.

15

We stood at the crater rim like ants teetering at the lip of a witch's cauldron. The heart of Bosavi steamed and bubbled below us. The volcano is long extinct, and the steams boiling up from below were merely mountain mists, infusing the trees within. The mountain itself stood above the clouds, a bleached white expanse stretching out towards the horizon in every direction. Beneath the cloud the forests breathed and a billion creatures, known and unknown, stirred with the sunrise. And there below was also my prize: a forgotten world shut off from the outside through time, unexplored by scientists or adventurers, unknown to all but the tribes that live there. My next steps would carry me down into the crater, into the lost domain inside the volcano.

Fogomayu's open airstrip offered an uncustomary view to the horizon, and that evening as the clouds cleared we had our first view of Bosavi. The dark massif stood proud from the forests in jagged grey silhouette, appearing to have two peaks. This is just one of Bosavi's many faces. On good aerial maps you can see the volcano's crater as the most perfect circle imaginable, with the highest points of the rim being over two thousand metres high. New Guinea's geology is rather unusual. I like Tim Flannery's description:

> New Guinea sprawls like a vast prehistoric bird across the sea north of Australia. After Greenland it is the world's largest island, and its size, shape and rugged mountains are all the result of its particular geological history, for New Guinea is Australia's bow wave. As the continent of Australia has drifted north, it has accumulated islands and fragments of other continents along its leading edge. Like dust swept up by a broom, these have built up into a long chaotic line of landforms.

Throughout the long period of continental drift that Flannery describes, great volcanic belts were formed along this bow wave. These gave rise to massive eruptions and created many of the mountains running up to New Britain and the Solomon Islands. The compression of New Guinea formed its mountainous spine, which in some places reaches nearly five thousand metres. This geology is one of the most important factors in the island's biology. Moisture-laden winds race inland and are driven higher by the central spine, where they cool to form the heavy rainclouds that make this one of the wettest places on Earth. The rainfall combines with the volcanic soils to make for ludicrously lush ecosystems. Volcanic soils are extremely fertile, because as they weather and break down they release phosphorus, iron, magnesium, potassium, nitrogen and trace elements, which have previously been locked up in the molten rocks. They are all essential to the growth of plants.

Many ecosystems are determined and defined by their altitude. Thus sea-level swamps are filled with mangroves, reptiles, wading and diving birds. Then as you ascend just a few hundred metres, moving through cloud forest and alpine meadows, the fauna changes dramatically until you reach the highest peaks, which are populated only by plants and animals that are pure altitude specialists. These mountains are incredibly important, as the flora and fauna that survive there are as good as stranded on islands, surrounded by oceans of lowland forest. They cannot survive in the different conditions found below, and are often trapped up there in the clouds. This leads to extremely high levels of endemism on these mountains (that is, there are huge numbers of animals that occur here, and often nowhere else on Earth).

Because of needing to recover from my fall, I'd had no part in the planning and recceing of Bosavi Mountain and was not over-optimistic about the outcome. I was disappointed that there was to be no death-defying rope work, no abseiling out of helicopters or bungee-jumping in on jungle vines. If it was merely a walk in, why did they need me? My role in these expeditions is supposed to be of 'the naturalist who gets into places others can't reach'! If I couldn't fulfil this role, then I'd be surplus to requirements.

It is of course nonsensical that it should be this way in my head, but it relates to something that's hardwired into my psyche. It started with my parents' entirely legitimate maxim: 'The harder you work for something, the sweeter the reward will be at the end.' And what could be truer? A squished sandwich eaten in the rain at the end of a massive rainforest yomp tastes more sublime than the finest haute cuisine. When you witness a sunrise on top of a peak, alone and fighting to find breath, it's particularly special – and it's an experience that creates more of a sense of being alive, more spiritual ecstasy, than half the world will ever know. Somewhere along the line, though, my mind seems to have taken this perfectly reasonable truth and twisted it into: 'Nothing is worth doing unless you are nearly killed with the effort of achieving it.'

This probably seems absurd to many readers, but not only is it quite a common human trait, it is becoming more and more prevalent. In my adult life I've very much fallen into the adventure community, climbing big mountains, doing fell runs, adventure races, ultra-marathons, triathlons and so on, and at every event will inevitably meet a score of people cast from the same deeply unattractive mould. They are obsessive, uber-competitive, stern, and sneer at anything as paltry as a mere marathon, or that might actually involve enjoyment. They pride themselves on not suffering fools lightly, take themselves incredibly seriously, pour scorn on everyone else's achievements, and will not do *anything* that is not utterly hardcore.

A perfect example of this personality is exemplified in *Eiger Dreams* by Jon Krakauer (who also wrote *Into Thin Air*, about his ascent of Everest and the ensuing tragedy that killed nine people). Krakauer finds himself in the climbing mecca, Chamonix, and gets talking to an obviously able young French mountaineer:

> He allows with characteristic Gallic modesty that he is both an expert parapente pilot and a *superbe* rock climber. I respond that I too happen to be a climber, and that I've been quite pleased with the quality of the routes that I've achieved in Chamonix thus far. I go on to tell him that I especially enjoyed the route I'd done the day before, a classic test piece rated *extrêmement difficile* on a slender, improbably steep spire called the Grand Capuchin.
>
> 'The Capuchin,' replies Patrick, clearly impressed, 'that must have been a very difficult summit to launch the parapente from, yes?'
>
> No, no I interject, I simply climbed the peak, I didn't mean to suggest I'd also flown off it. 'Non?' says Patrick, momentarily taken aback. 'Well, to

solo the Capuchin, that is a worthwhile undertaking all the same.' Actually, I sheepishly explain, I hadn't soloed the peak either, I'd done it with a partner and a rope. 'You did not solo and you did not fly?' asks the Frenchman, incredulous. 'Did you not find the experience ... how do you say in English ... banal?'

I have experienced the equivalent of this conversation so many times, it makes my brain bleed thinking about it. One memorable one was in the Blue Mountains near Sydney, talking about a trip to a canyon known as Claustral: a big day out, and one of the most magnificent adventures to be had in Australia. I was in a bar chatting about it to a muscular Norwegian girl who looked as if she could do one-arm chin-ups while carrying me under the other arm. She sniffed in utter contempt: 'I have done Claustral so many times, I just find it boring.'

I find these people unbearable, yet recognise all of these instincts in myself. At least, though, if I can acknowledge them, hopefully I can do my best to battle them – most of all not to take myself too seriously – surely one of the least appealing of all human traits. Bosavi, though, was waking all of these instincts in me.

We would be dropped off on the lower slopes by helicopter, and merely walk in over three or four days. It was all I could do not to interrupt our planning meeting by screwing up the map and sneering: 'I have done expeditions like this so many times, I just find them boring!' This trip was supposed to result in an iconic adventure film offering jaw-dropping feats and challenges, taking us all to the limits and back again. To make matters worse, from the photos I'd seen, the four-kilometre-wide cauldron of Bosavi was just too big for you to appreciate; it was a perfect crater and would probably just look like a big wooded hill. That, and my by now deep belief in the curse of New Guinea, meant I had set my heart on being underwhelmed by Bosavi. How wrong I was ...

Our first recce flight was one of the most spectacular helicopter trips of my life. From Fogomayu at a little over sea level, we flew steadily upwards over the mountain's perfect green lower slopes, before it became abruptly steeper and suddenly we could see the ridges of the outer caldera. One of these ridgelines leading to the summit had a path running up it. Were you to stray from it and drop down into one of the gorges at either side, you would be in some of the most impenetrable and unforgiving terrain in the world. As we neared the rim, the outer slopes got even steeper. Then we swooped up to the edge, which appeared to be no more than a few feet wide. As the helicopter roared over it, suddenly there was empty air below us, the inside of the cauldron dropping a thousand metres down. From up high, the caldera appeared perfectly smooth, but as the heli began to drop, the near-vertical walls towering above us, we spotted potential weak points that we might be able to exploit. If we were going to descend into the depths of Bosavi, one of the ridges running down to the bottom would have to be our route.

The bowl is four kilometres in diameter. Wherever it is not vertiginous rock it is densely forested. It contains manifold peaks, rockfaces, valleys and raging watercourses. A unique and complex landscape with a billion hiding places, it was large enough to envelop a decent-sized town. It would take a lifetime to explore it properly. Trees that had looked like saplings from up high were actually massive hardwoods. Streams that had looked like trickles we now realised were in fact thundering rivers. When going over the maps back at basecamp I'd fancied our chances of following a streamway down into the bowl, but even from the helicopter I could see this would be folly. A further challenge was that while Luke, Tim and I would be making our way on foot down into the crater, the rest of the crew would have to follow later in the heli, which meant we needed to find somewhere that could function as a helipad – and a helipad needs probably fifteen square metres of open, flat terrain, of which there was none.

This small crew had been chosen because of my strong belief that on potentially dangerous missions you're better off with fewer participants. Every person needs to be self-sufficient, and any extra you add is just another person to feed, find shelter for and risk getting hurt. Tim was along to keep an eye on safety, and on Luke. Our prime hard-nut cameramen Johnny Rogers and Keith Partridge had been counted out because of illness and injury, so Lucky was to get his crack at proving himself tough enough to endure one of the most challenging terrains imaginable.

Fuel is always the factor that can bring recce flights to a halt, and our pilot had his eyes firmly on the fuel gauge. Just as it looked as if we'd have to call it a day, something caught his eye down below. We had been hovering over quite a major river, looking for a flat or rocky bank that would meet our requirements. What caught his attention, though, was a stream tumbling down a rocky slope in a series of waterfalls. At some stage there had clearly been a flash flood, which had brought down tonnes of mud and rock and rent the hillside clear of vegetation. What was left was a semi-detached-house-sized heap. Perhaps if we could get down there with machetes we could get rid of the trees and shrubs and turn it into a helipad. The pilot was eventually in agreement: 'You'd need to flatten it all out, and clear down all those trees, but then it might work – rather you than me getting down here, though. SHIT, man!'

He got us down as low as he dared, hovering over the mud heap long enough for me to programme the spot into my GPS. There was another landmark in the form of a great white tree at the base of the landslide. It remained etched in all our memories as 'the spirit tree', the spot we'd be aiming for.

Still watching his fuel gauge, the pilot eased back on the controls and the heli rose up towards the rim, then beyond to the outer slopes of Bosavi. There was a clear spot on one of the lower ridges, so he dropped us there, wishing us luck before flying off and leaving us in eerie silence.

As we waited for him to return with Maxi and Chocol, a New Guinea harpy eagle flew overhead with its penetrating 'bonk bonk' call. The sound echoed hauntingly around the silent gullies, emphasising the quiet. The New Guinea harpy is one of the largest eagles in existence, very rare and seldom seen. It looks almost identical to the (unrelated) harpy eagles of Latin America and the Philippines, with its light breast feathers, dark wings and head, and crown of feathers behind the head. Harpies are the most powerful eagles in the world, and so large that much of their prey consists of arboreal mammals: in Latin America, monkeys and sloths, here in New Guinea cuscus and tree kangaroos, all speared in the tree-tops with their stiletto-like talons. The stabbing, delivered after a lightning-fast stoop or swoop, is so severe that when filming harpy eagles in Panama we had to climb into the tree-tops wearing stab jackets and police riot helmets!

This may seem like overkill, but one of my crew had been attacked by a fully grown female harpy; he hadn't heard her coming because the attack was totally silent, and he was hit from behind with incredible force. He believes that if he hadn't been wearing the stab jacket, the nine-kilo female would have punched her talons – longer than the claws of a grizzly bear – clean through his back, and probably killed him.

After the last chopper drop, we pressed on up the ridgeline along a tiny overgrown path. Maxi pointed out a bird of paradise *lek* (the place in the forest that they clear so as to dance for the ladies). This one was just an elevated branch in a clearing, picked clean of moss and vegetation; but some species will even clear a space in the canopy above so that a shaft of sunlight will spotlight them as they perform. Further up the path we came across the home of a Macgregor's bower-bird. This construction is not a nest, but a stage they build for themselves to dance around. It consists of sticks painstakingly arranged on a spindly maypole, and looks kind of like an old Christmas tree with all the leaves fallen off, or a half-finished game of KerPlunk. Around it a cleared patch of mossy ground decorated with shiny beetle-wing cases was the stage for our avian hero, who was fluttering around in the branches above us as we approached.

My greatest ever bird experience was with a different species of bowerbird in Australia – the satin bowerbird. His bower was an avenue of curved grasses, the front of which was adorned with anything blue he could find: flowers, butterfly wings, feathers, biro tops, drinking straws. I plucked a red flower from nearby and placed it in the bower. Then, while I sat there, no more than a metre away, he flew straight down, threw the flower out, and cawed angrily right in my face! Every time I put the flower back he'd come straight back and throw it out, more and more furious every time!

Bowerbirds are fairly unremarkable-looking birds. The satin bower-bird has glossy blue-black coloration and a vibrant lilac eye, but otherwise they're just quite nondescript thrush-like birds that occur only in New Guinea and Australia. The males are polygamous and mate with as many females as they possibly can; females cruise around the area looking for the finest bower they can find, and mate with the attendant male. Jared Diamond in his extraordinary work *The Rise and Fall of the Third Chimpanzee* notes: 'The human equivalent of such scenes is played out every night on Sunset Strip a few miles from my house in Los Angeles'! Diamond goes on to explain far better than I ever could the purpose of the bower in proving the fitness of a particular male and hence impressing the females:

> She knows at once that the male is strong, since the bower he assembled weighs hundreds of times his own weight, and since he had to drag some individual decorations half his weight from dozens of yards away. She knows he has the mechanical dexterity to weave hundreds of sticks into a hut, tower or walls. He must have a good brain to carry out the complex design correctly. He must have good vision and memory to search out the required hundreds of decorations in the jungle. He must be good at coping with life to have survived to the age of perfecting all those skills. He must be dominant over other males – since males spend much of their leisure time trying to wreck and steal from each other's bowers, only the best males end up with intact bowers and many decorations. Thus bower-building provides a comprehensive test of male genes. It is as if women put each of their suitors in sequence through a weight-lifting contest, sewing contest, chess tournament, eye test and boxing match, and finally went to bed with the winner! We can easily relate to this concept, just think of those adverts showing a young man presenting a diamond ring to a young woman. You cannot eat a diamond ring, but a woman knows that the gift of that ring tells of the resources her suitor commands and might devote to her and her offspring ... From this perspective bowerbirds are rather human ... we swathe ourselves in coloured cloths, spray or daub ourselves with perfumes, paints and powders, and augment our beauty with decorations ranging from jewels to sports cars.

So the Macgregor's bower is effectively just a big red Ferrari, being shown off to try and pull the ladies! Well, unfortunately the bird in New Guinea wasn't quite so evident or accommodating as the satin bowerbird had been in Oz. But I left a leaf there in the hope he would chuck it out before George came back a few days later.

After several hours slogging up the ridgeline, the greenery thinned out and then we were at the rim, looking out over one of the world's most spectacular views. All of my earlier cynicism acquired during the planning meetings evaporated. Rain clouds were gathering behind us, racing up from inside like billowing smoke. It looked as if the volcano was still active and smouldering, due another massive eruption after a hundred thousand years of sleep. As the clouds obscured the bowels of the crater, I felt an overpowering desire to just jump, to leap out on to those pillows of cotton wool and drift down into the Lost World beneath.

The next morning I would become the first outsider ever to venture into Bosavi. This place was truly overpowering, and what lay beneath we could only guess at. Here was an opportunity no one had ever had before, and no one would ever have again: to walk into this sacred world with the eyes of a child, discovering everything for the first time. I was suddenly overcome with emotion and felt like sitting down and having a cry. This was what it had all been about! This was a chance to lay to rest the ghost in my mental machine, the early expeditions that had left me broken and beaten. Now I had an opportunity to realise all the fantasies of my childhood. It sounds melodramatic, but right then I remembered the giant Papuan stick insects I had gazed at in awe in those glass cases as a kid, every misty pith-helmeted machete-wielding fantasy, every book I'd ever read on Papua, and I *really* wanted to be the first person into that crater more than I've ever wanted anything in my life.

After the failure of Irian Jaya I had filed away New Guinea in the back of my brain, but now I could make it all right. I sat there shaking, until the rains drove me to seek cover.

Of course, I have to make clear that I would only be the first *outsider* to enter the crater. Bosavi and its surrounding lands are owned by several clans of Papuan people who live on the outer slopes of the mountain. They see the mountain as sacred, and have to speak hidden magic before they can venture inside. As Chocol had told us, they only go in every few years, if they have a big festival, because – in their own words – it is so full of animals to hunt. Because it is such a spiritual place for the clans, brothers Maxi and Chocol were to accompany us into the cauldron, though they'd certainly not function as guides, as neither had been inside before.

When cameraman Lucky Luke, Tim Fogg and I reached the summit ridge, we were joined by ten other men and women who had all come to join in the ceremonies that it was essential to hold before we could go any further. It was a bit of a tricky time. In New Guinea *tok tok*, discussion, is vitally important to every bit of business that takes place. It can last for days, and is particularly essential when sending outsiders with all their precious 'stuffs' into the most sacred place. It was difficult for us, as we had such limited time. We were totally self-sufficient as far as food, fuel and shelter went and were carrying everything on our backs – we certainly didn't have enough food to take even one extra person. Additionally, if any of the other locals decided they wanted to join us, we would be putting ourselves in the awkward position of being responsible for their safety.

Our diplomacy continued through the afternoon and into the evening. Every possible option was discussed a thousand times, with tutting, shaking of heads and vociferous pointing of fingers. As the light disappeared and the rain thundered down on the tarp above us, the firelight illuminated the fabulous faces visible in the smoke. Maxi and Chocol's wizened mother fed the fire with dried sap from a lowland tree. The white smoke had a scent more sumptuous than any incense I've savoured.

Their father Seguro was the Big Man, and clearly demanded attention when he spoke, shouting everyone else down. He had an aggressive, slightly manic streak that made Tim and me exchange knowing looks across the smoke. This was not the kind of person we'd want with us on a dangerous journey. As Seguro spoke he would leap to his feet and gesticulate wildly, the light from the fire glinting through the hole in his septum – on ceremonial occasions he would have worn a bone, a carved nautilus shell or perhaps a ballpoint pen through it. He was a magic man and a storyteller, and as he leapt about in the flamelight he would become now a great hunter drawing back his bow, now the tree kangaroo running for his life, now the great snake that inhabited the high peaks of Bosavi.

Eventually, late in the evening, Maxi interpreted again for Seguro, telling us that he had given us his blessing to enter the crater, and tomorrow would sing the special magic that would open the mountain and prevent 'nature' being angry with us. We needed to respond in some way to their stories, so Tim pulled out his tin whistle and played some beautiful, wistful Celtic tunes. It's such a simple instrument, little more than a tin tube with a plastic mouthpiece, yet the clear, piercing notes instantly transport you to another time and place. I find it utterly haunting and it always brings a lump to my throat. The assembled Sienna Falls folk, however, talked all the way through his tunes! This really raised my hackles. I wanted to tell them to be quiet, to show Tim the same respect we'd shown them, but eventually I realised that this was their way of accepting his contribution, of making it a part of the *tok tok*, and a subject for – well, talk-talk. When he'd finished they all wanted to know what the story behind his tunes was, and seemed disappointed when Tim's answer didn't take the rest of the night.

In the early hours of the morning we walked back up through the soggy, mossy forest to our hammocks. As we wandered along the rim, my torchlight startled perching birds, blinking as we passed. The sounds of frogs resounded amongst the ferns. I spooked a tiny sleeping bird, no bigger than a wren. It landed a foot in front of my face, clearly bewildered. Luke was approaching behind me and I tried to stop him, but he nudged the branch. The bird took off and landed on his head! We stood there stupefied for several minutes. The bird wasn't showing any signs of going anywhere, the camera was nowhere near, but it was worth a try. As I slowly made to fetch it, though, the bird took off in panic and didn't return.

As we nestled into our hammocks and the enveloping comfort of our sleeping bags, the sky at the western horizon was burning orange with the flames from roaring oil wells, foreign multinationals raping the land and encroaching ever closer on what remains of this extraordinarily precious wilderness. My mind danced with visions of what we might find beneath: forgotten dinosaurs, bizarre new species of mammals, perhaps the last ever Tasmanian tigers or super-sized versions of already weird Papuan creatures. To sleep perchance to dream . . .

Next morning we emerged just before first light into a landscape of spindly trees, gift-wrapped in green: vegetation festooned with bryophytes – mosses and liverworts, ferns and tree ferns. The montane plant life of New Guinea is one of the most intriguing in the world. These peaks may well be the last refuge of the flora found throughout Australasia and in South America. The ancient megacontinent of Gondwana stretched from the South Pole to the Equator, and would have spanned the southern hemisphere that now comprises Africa,

Madagascar, South America, Australia, New Zealand and New Guinea itself. This vast landmass would once have had common plant and animal life, but as the continents began to break up about 167 million years ago, climatic conditions changed, alternately scorching and freezing the earth, and land bridges rose and fell with sea levels and glaciation. Living things would have surged and invaded, then separated and splintered.

This green wonderland that I was now surveying could be a window back into prehistory, a last glimpse of the Jurassic period, when Gondwana started to break up. For scientists, particularly palaeobotanists – who piece together the puzzles of prehistory through plants both modern and fossilised – these places are an endless mine of information and pose never-ending questions. Did Antarctica once house hot, green tropical forests? When did the main climate upheavals take place, when did each creature and plant become extinct? How did those ancient landmasses connect, then drift apart?

The walk down to that crater rim was like looking back through time. Though the flora of New Guinea is still poorly studied and understood, it is probable that many of the species I found around me now were trodden upon by mighty dinosaurs, the first mammals, and the first early precursors of man ... seething tendrils of cloud billow up the ravaged walls, as if witches are mixing some primeval potion in a giant green cauldron. Out beyond the mountain and off into the distance is a sea of cloud, with only a few taller peaks poking up through the perfect unbroken white blanket. Above the fierce whiteness is all blue skies and a quickly rising sun. This is what is known as an inversion; where the usual thermocline is reversed, and cold air lies in the valleys with a layer of warm air above it. Cloud condenses here, and forms an ocean of cloud. Often this will burn off after a few hours when the sun starts to redress the balance and air layers start to mix. Tiny honeyeater and colourful fruit doves perch on the moss-covered boughs about us, chirping inquisitively at these strange white intruders. Even my cynical godless Western mind can see why this place is seen as having such powerful magic to the local people. To further emphasise the point, Segoro sings us into the crater, creating a gate using a magic stick. He chants a refrain to the Gods of Bosavi crater, to a giant serpent said to dwell around the rim, and to the tree kangaroos within, imploring that they give our journey their blessing. I can feel the thumping in the back of my throat as we cross under the gateway.

We begin our descent. The crater wall drops steeply like the fin on a spaceship. The ridgeline is thickly forested, with dense undergrowth, and some fair-sized trees as thick as my thigh hanging on for dear life. The quickest descent would be a precipitous butt slide through the bushes, but by zigzagging we manage to maintain a semblance of control. Some sections are slippery vertical rock and can only be negotiated using roots as ropes. It's a war of nerves, entrusting your whole body weight (plus a backpack that weighs as much as a couple of crates of lager) to a spindly piece of plant, but after a few hours with the rain pounding down on us we are indifferent to everything, alone in our own world of misery.

Something new and surprising is that for the first time since arriving in New Guinea we find ourselves covered in leeches. At one stop, I count sixteen on one boot. They are generally small and individually insignificant, but they occur in substantial numbers, and every now and again one will latch on and make a real mess. Most haemophagus (blood-drinking) leeches have some sort of anaesthetic in their saliva so that the pain of the bite is reduced or not felt at all, but you do feel some of the larger species getting stuck in when their teeth cut into your flesh. These little ones we encounter, though, are painless, and it's not until you find them, fat and happy, guzzling your blood that you know they're around. An anticoagulant in the leech's saliva means the blood won't immediately stop flowing after they've had their fill and dropped off. At a rest stop, after a couple of hours, we take stock and rip off as many as we can find. Luke has chosen to walk in knee-length board shorts, and has, not surprisingly, suffered rather worse than Tim and me.

I spot one busily at it above Luke's collarline. 'Lucky, you've got one on your neck, buddy.' He rubs his neck and it falls off, but the wound continues bleeding profusely. Sometimes this happens – the anti-coagulant has kicked in and the wounds gush. 'Mate, it looks like you've been bitten by a vampire!'

'Stevie Boy, you've got a bit of a blood stain on your shirt.' Luke points to my stomach. I lift my shirt to reveal two curled up around my belly button. Maxi and Chocol idly pick theirs off themselves and flick them into the undergrowth.

'Oh well, boys,' I comment hopefully. 'Lots of leeches does of course mean lots of animals that they're feeding on – that's good news for us.'

'Well, unless they're the kind of leeches that mostly feed on frogs or snails or something,' says Lucky.

'Well, no, usually they stick to one kind of food. I'll bet these all feed on warm-blooded animals,' I tell him, ever the optimist.

Luke has taken up a bottle of bug spray and is blasting the leeches on his ankles, which wrinkle up like salted slugs and drop off. 'That's a bad idea, Lucky,' Tim tells him. 'That stuff kills them, but it makes them vomit their stomach contents into the wound – it's much more likely to get infected like that.'

'Yar bru, but it's really satisfying seeing the little bawstards die this way!'

'Just spray it round your laces and the top of your bootline, Luke,' Tim advises. 'That stops them getting into your boots.'

I investigate a squelchy sensation in my right boot to find one bloated leech lying inside my sock, probably saying to himself, 'O-o-o-ohhhh no-o-o-o, I'm so full, I couldn't eat another thing.' The back of my heel looks as though someone's hacked it with a machete. It's truly incredible that a tiny worm could cause such devastation, particularly as the actual bite mark is no more than a few millimetres across.

'Dis leechie love white pella taste,' giggles Maxi. 'You very sweet to eat!'

As Luke continues the scrutiny, he unzips his board shorts, looks down where no man wants to ever find a leech, and lets out a groan. 'Aw no, Steve buddy, I've got one in the honeymoon suite!'

'Well, what do you want *me* to do about it? I'm not getting them off for you!'

'I didn't ask you to bru – aw Jeez, they're drinking away down there – I can hear glugging!'

'So why do they call you Lucky – Lucky?'

'Now Mister Luke got two little kok!'! Maxi adds helpfully before collapsing into hee-hee-hees.

Luke pulls his hand out of his shorts, covered with blood. 'Oh bru – this is not good. Not the honeymoon suite . . .'

We have ropes with us to tie up handlines if anything should get too vertical, but the truth is it's all pretty vertical, and to rope up every section would take us weeks. Maxi is sliding away in front of us, only using one hand, as he's carrying our axe in the other. I offer to take it for some of the way but he won't hear of it. I have a feeling that he has his eye on it for himself and is scared that one of us will steal it from him.

After a few hours the rain really starts to hammer down. Drenched to the skin, we have to keep on moving just to stay warm. The way down is remorseless, and we can't relax for even a second, as one lapse in concentration could result in a slither that could be terminal. There are more stinging nettles here than I have ever seen anywhere, and vicious spined trees and rattans too. After just a few hours our hands are full of splinters, sliced by sharp stems and rubbed raw from clinging to roots. They're agony, and I'm going to have to work hard to make sure they don't get infected.

Every once in a while we stop on a ledge, and look across through the constantly shifting clouds. By now we were well down in the crater, its rock walls towering behind us. We could only very rarely see right across the bowl to the other side, and it seemed likely that these boiling clouds would be our companions for much of our time in here. We came to more and more bare, slippery vertical rock, and had to work together to lower each other down on to them. Usually I went first, Luke passed the camera down, then placed his foot on to a step made with my hands, Tim helping from above. Sporadic bursts of torrential rain made it incredibly difficult for Luke to film and to keep the camera dry.

It's one of those impossible situations where the harder things get, and the more difficult it becomes to get the camera out, the more you want to be filming non-stop. It's a real struggle for Luke – the truth is that every scrabble down a rockface is worth filming, and every new view is a better picture than anything we've seen in our month or so at basecamp. Ideally you'd want to film every second of every step you take further down into this Lost World, but you just can't. The camera would be useless within minutes if it was kept out in the rain. And if we had to wait at every slide in order to position Luke at the top, then the bottom, we'd be weeks getting down into the crater. Also, at the forefront of Tim's and my minds was that we had another job to do here, not just make a film. We needed to get off this ridgeline as soon as possible. It was lethal, there would be nowhere here to make a camp, and we only had enough provisions for four days. If we didn't get down, find and build our helipad in that time, we were in for a very rough time indeed.

Early afternoon, after about seven miserable hours on the go, and we broke out on to a clear, bubbling creek with a large fallen tree spanning it – a perfect seat. We sat out in the rain and broke open our lunch.

'So what's the scran for today then, buddy?' Luke asks, as if expecting me to give him a menu.

'Don't ask, Lucky. We have two crackers each for lunch, each day, with a jar of peanut butter to share between the five of us – for four days.'

'That's just for appetiser, right? I'm a growing man, Stevie boy.'

'Except in the honeymoon suite, eh Lucky?'

'I reckon we've broken the back of the steepest section,' Tim says. 'Pretty soon we're going to have to take the plunge, break from this ridgeline and start to find our way to the chopper landing site.'

'What does the GPS say, Stevie?' Luke asks, while scraping a carefully measured thumbnail of peanut butter on to his Lik Lik Wopa (Little Whopper) cracker. Fried in beef dripping, these chunky biscuits taste a bit like doormats, and instantly suck every ounce of moisture from your body. The peanut butter doesn't help this. I fight to speak through a parched mouth and past my swollen tongue.

'Looking at the GPS reading taken from the chopper, we're less than a kilometre away as the crow flies from the landslide, which is a positive start.'

'*One k?*' Lucky is ecstatic. 'That's awesome, we'll be there in an hour!'

'Hmmm, I don't think so, Lucky.' I hate to crush his optimism. 'We've been going seven hours from the ridgeline and only gone eight hundred metres!'

'You're having an actual laugh there, buddy, I'm dead already.' This realisation takes the light out of Lucky's blue eyes. 'We could be going days before we find the place, then.'

'Well, look, we have food for four days,' reasons Tim. 'We don't have to push it too hard today. Let's give it another few hours, and if a good camp site turns up we'll crash there. For now, though, we need to figure out a route.'

'Hey, Chocol buddy, that peanut butter has to last us four days.' Luke's eyes narrow. Chocol has taken about a third of the jar on one single Lik Lik Wopa cracker. He looks up innocently like a child who's been caught with his fingers in the cookie jar.

I looked downstream at the lovely clear water. 'My *instinct* would be to follow the stream bed – chances are it will eventually hook up with the river that flows under the landslide. And if we hadn't had such a good helicopter recce, that's exactly what I'd do.'

Tim: 'Yeah, but that would be a total disaster. We could go round that corner and end up at a vertical waterfall or get stuck in a gorge. Don't forget we probably still have more than a hundred metres of vertical descent to do. I reckon we'd be better to follow the contours of the land, cutting a trail round parallel to the main stream we hope the landing site's on.'

There was no map to go on, no view now that we had dropped down into the bowels of Bosavi. We were bound to instinct and our GPS, knowing that the reading we had was probably far from accurate. I was feeling the weight of responsibility, and kind of wished I'd given Tim the GPS from the start. He's better at these things than I am.

We marched hard until past two o'clock in the afternoon, then three, then four, thrashing hour after hour up and down some of the meanest terrain you will ever see. A high ridgeline lay to the south of us, cutting us off from where we assumed the main river had to be. Late afternoon, and Luke was getting really tired. There seemed to be less life in his normally sparkling blue eyes, and he was stopping more and more often for longer and longer breaks. Though he wasn't carrying a load, it was a hell of a demand, expecting him to keep up with us all while filming everything. As I've already mentioned, despite being in his fifties Tim is the most effortlessly fit and tough person I know, but even he was starting to look a bit grey and haggard. The stress of continually risking your life and your friend's life on a flimsy root handhold can be fiercely draining, and everyone's nerves were shot. The next time the sky was visible, we stopped for me to take another reading.

'What's the good news there, Stevie boy?' Luke asked hopefully.

I looked at the results and winced. For a second it crossed my mind to lie through my teeth, but we were all adults, and we were in this together.

'Well, Lucky, according to this, we're now 1.2 kilometres away from the site—'

'*What?* How can we have got further away?!'

'It gets worse. We're also a hundred metres higher up than we were at lunch.'

Luke sat down and put his head in his hands. It was a bit of a low point for all of us. We were exhausted, hungry, torn to shreds from slithering and from the sharp, spiky plants, soaking wet, and cold. On the positive side for me at least, my ruined ankle was holding up pretty well even with the terrain and my huge pack. Admittedly, I was demolishing painkillers like they were sweeties, but I'd been yomping on ahead of everyone, even Chocol and Maxi, and having to wait in the rain for everyone to catch up every ten or fifteen minutes.

'We're going to have to break that other ankle for you, bru,' Luke jested.

With only a few hours of daylight left to go, we crashed out of the thick forest, and there was the largest stream we'd come across so far, big enough in fact to even qualify as a river. Luke and Chocol were exhausted and couldn't go any further, so we concocted a quick battle plan. Neither of them would be able to do much in terms of setting up camp, so Tim decided to stay behind and take on that job while Maxi and I pushed ahead to try and get a better fix on our location. We ditched our packs, taking nothing but a machete each, and followed the river down to where I was sure it would meet the other river where we'd intended to temporarily camp. After a ten-hour day carrying thirty-odd kilos on my back, it felt as if I was walking on air.

However, after forty minutes of slipping and teetering down the increasingly torrential flow, the deep tiredness in my bones was starting to take its toll. After a few careless slips that nearly sent me over bone-breaking drops, I forced myself to concentrate hard on every step. Just when it seemed we must surely meet the spirit-tree river, we hit the top of a twenty-metre waterfall coursing down a twisting crack in the rock then over what appeared to be an overhang down into a raging plunge pool. The whole thing was hemmed in on all sides by impenetrable rock walls, and we couldn't see down to what lay below. As we approached, a cloud of giant flying foxes that were roosting above it took to the skies, filling the narrow gorge with huge flapping eagle shapes, panicking as they tried to escape the first humans they had ever seen.

It was a dead end. We could have tried to fix ropes to abseil down it, but if the pool below offered no easy escape, this would be suicide. Though we couldn't see well what was below, there would be no way on earth we could get back up those ropes. Worse than that, when we listened carefully to the sounds of the streamway, beneath the constant roar of the raging water was a heavy, dull clocking sound, the noise of beachball-sized boulders rattling around as they were dragged down the flow. These might not dislodge very often, but if one plunged over the edge at the moment one of us was descending, no helmet would make any difference. It'd be instant death.

'This ples he go down no good. Down der is big buggarap, Mister Steve. We go der is dai pinis. Tut-tut-tut,' Maxi notes, blinking his milky eyes and twirling his fingers contemplatively in his beard.

By now the sun had gone down and the forests were starting to get dark. We backtracked fifteen minutes or so, then climbed up through the scrub to a high point where we hoped to get a view out to the south. In the early days before GPS, cartographers and explorers would map the jungles and find their route by climbing to the tallest place available, to get a vantage point over the land around them. In these forests, which hem you in under a duvet of green, this is all but impossible. But here we were, deep within Bosavi crater and having to do just that in order to find our way. All we had as a visual reference point was the memory of the landslide, our white spirit tree and the small waterfalls we'd seen from the air. So Maxi and I scrambled up a ridgeline to its highest point.

As we neared the top the ground steepened and the foliage thickened, so that we had to hack through the tearing vines and thorny branches with our machetes, using the roots as handholds to drag ourselves upwards. By this stage we were so hacked and scratched that I no longer cared too much about the thorns, and just dragged myself through the bushes with brute strength, ripping my T-shirt and exposed skin even more. It had been a heavy, frustrating day, and I'd about had enough.

At the top of the peak, we could just make out a vertical drop of perhaps a hundred metres. We must surely be able to look down and see our camp site below – but even there the vegetation was just too thick. There was, however, a spindly tree. Maxi sprang straight into the branches, clambering upwards like a chimpanzee. Many of the people I've met in New Guinea have unusually shaped toes – a huge big toe three times the size of mine, with a large V-shaped gap between it and the next toe. Believe it or not, men like Maxi can actually grip boughs between these toes when climbing up vertical trunks, hand over hand. He is extraordinarily adept, but seeing him now, standing twenty feet up the swaying tree, knowing that if it toppled southwards he'd drop off into nothingness, I vowed to never tell the risk-assessment crew about this one.

Determined not to be outdone by Maxi, I tried to haul myself up another nearby tree, but my extra bulk bent it instantly and it started swaying as if to drop me over the edge. I scrambled down, my heart racing, and made an executive decision to leave the surveying to the experts.

'Can you see the landslide, Maxi? Can you see anything at all?'

'No can see im. I go up some more.' Maxi goes hand over hand even higher. The tree was now no wider than my bicep and was lurching from side to side. I feel slightly sick.

'Mister Istiv, mister Istiv!' Maxi shouts down to me, excited. 'This white *diwai* . . . the tree mister Istiv. I see him! I see him tru!'

Could this really be the white tree I'd seen from the helicopter? Could we have finally got a fix on our destination? The sky above me is relatively clear so I take out the GPS and line up the arrow that tells where we should be aiming for. It registers the exact direction that Maxi is pointing in, and says we have no more than two hundred metres to travel. I could have leapt up and down for joy.

By the time we scrabbled down to the riverbed, bruised, battered and utterly spent, Tim had set up quite a decent camp. Chocol had chosen to build a local-style shelter for himself and his brother, a low stick roof covered with broad leaves and ferns to provide some degree of waterproofing. The floor was laid thick with ferns and a fire was already burning inside to keep the mosquitoes at bay. We clustered around the flames, as the rain was doing little to keep the ravenous critters at bay. After such a brutal day we looked a right sorry bunch, our clothing torn and sodden, brown and green with mud and plant goo. Chocol was so tired he could barely lift his shiny bald head from his fern bed, and just lay there in the green poncho we'd given him. Luke looked as if he'd been stabbed in the throat, with leech blood all over it and staining his shirt. Tim's hands and forearms were like mine, a mess of scratches, cuts and thorns. We looked beaten.

'So what's the score, Stevie? Give us the good news.'

I don't want to give them false hopes, so I start carefully. 'Well, we're close, boys, but this riverbed is no good, it ends at the top of a waterfall forty minutes or so that way . . . Hard going, too.'

'Can we abseil the fall, Stevie?' Tim asks.

'Not with the kit we have now, Tim, and it'd be a real drop into the unknown anyhow. Abbing *inside* the waterfall with boulders pitching down at your head, and if you got to the bottom and it was no go, you'd never be able to climb back up. Sketchy, to say the least.'

'So we're rooted!' Luke groans. 'What happens now, bru?'

'Well, no, there *is* one other possibility. We got to the top of this peak here' – I point in the direction of our high viewing point. 'Maxi got up a tree and looked down into the next valley, and he reckons he could see the white tree.'

Both Luke and Tim sit up rigid at the mention of our fabled talisman.

'It's about four hundred metres in that direction, and I reckon we can skirt the peak and try and find a break in the cliff. Either that, or we just push on downstream until the cliff ends.'

'I'm guessing this is a job for the morning, eh?' Luke sounded seriously worried.

'Hell yes! I think we should leave the camp in place tomorrow and try and make a route through. Then we can come back for our stuff, or worst-case scenario, camp here again tomorrow night. It's nice enough here, isn't it?'

Nice was overstating it. But in fact it was the only place we could have made a camp site so far, with decent-sized trees for the three of us to hang our hammocks from, fresh clear water, and access to the sky for our GPS, though nothing like enough to make a satellite phone call. The only real issue was going to be the constant roar of the river, which would inevitably make sleeping tough. That, and the ever-present threat of a flash flood.

'Right, that's sorted then. A good night's sleep, then up at sparrow's fart and get cracking. So Timmy, what delights do you have for our dinner?'

'Trust me, Stevie, you don't want to know ...'

Luckily, we didn't have heavy rainfall that night, and the water level had actually dropped by morning. After a measly breakfast, we stowed our packs except for the ropes and machetes, and set off in the direction of our GPS arrow.

The first surprise was hitting a flood plain caused by our own river – the first piece of flat ground we'd seen within the crater. It was probably around fifty metres by thirty, and was for Tim and me a major discovery. Although it was densely forested it was flat, which meant that at the very least we'd be able to build a camp to provide a home for a minimum of fifteen people here. Its position near the water would be a really good thing, though again the threat of a flash flood would be a worry. We made a mental note of it, and kept on moving.

The vegetation here wasn't as dense as on most of the journey down, and we could carve out a path without having to hack every step with our machetes. This meant we got to the brink of the drop-off within no more than twenty minutes. This was our first stroke of luck. The second occurred at the drop-off itself. On later exploration of this cliff we found it to be unbroken, vertical and impassable for a considerable distance in both directions – except in one place, right here. We were at the top of a chute where the slope wasn't vertical. Soil had been laid down over the years, enabling hardy trees to plunge their roots in and suck a life from the ground. The chances of us getting so lucky were negligible, as you couldn't see more than ten metres in any direction. But the gods were truly with us that morning.

The slide down, though, was a dangerous one, protected only by the roots and trees themselves. Saplings as spindly as walking sticks would quite comfortably take even my weight, whereas some irritating palms that were as thick as a lamp-post would collapse, as if rotten right through. Plenty of trees protect themselves with barbaric spines, and they'd inevitably be the ones you'd grab out for in moments of panic.

Less than an hour after we'd departed from the previous night's camp we dropped down to the gorge beneath and waded into the river. This was a proper river. It was as wide as a tennis court is long, and flowed with considerable power, power that could obviously increase dramatically, from the evidence of several giant trees lying in the streamway. The water was perfectly clear, and there were some glorious azure pools, maybe four metres deep. To the north (from where we had just slid) was a rock wall perhaps eighty metres high with forest above, and to the south another rock wall, about half as high, with forest going up and up and up until it met the cliffs that extended to the summit. A thousand metres above us, the inner walls of the caldera seemed to hang right over our heads, increasing the sense of majesty. The river flowed from east to west, and downstream of us rushed into a tight gorge. The water there was too deep to wade through, with no room on either side to skirt the flow, so our only way on was upstream.

Given the extreme effort of getting down here and the terrain's natural protection, it seemed highly unlikely that even locals would ever have found their way down. We had no doubts that every footstep was the first any human had ever taken here. And we had the extraordinary privilege of seeing this mind-bendingly beautiful place for the first time, of bringing it to the outside world. This was real, genuine old-fashioned exploration. As if we were wandering through a mighty cathedral, no one dared speak, so great was our reverence for what we were experiencing.

Ecstasy was zinging through my veins like an overdose of adrenalin, hairs standing up on the back of my neck and forearms. Surely just up ahead would be our destination? Would it be a suitable landing site? Would we be able to build a camp there? If the answer to any of these questions was no, we were in deep trouble.

The gorge was startling, so wondrous that we just wanted to soak in every second and imprint it on our minds for ever. We were wading in bubbling, glassy water. The tangled jungle canopy above blocked out the sun, and huge ferns and spindly palms drooping down around me made the gorge feel like a film set of the Cretaceous period. I'd barely have batted an eyelid if I'd looked up to find a foot-long dragonfly hovering around my head. We swam through malachite and jade pools as we pushed upstream, occasionally teetering over mossy fallen tree trunks that spanned the waters. Then further and further into the unknown. At last, there it was, the most beautiful tree I have ever seen, a colossal white spirit trunk glowering over a tumble of landslip mud and rocks. Behind the landslide was a rock wall with a small waterfall tumbling down it.

We'd made it.

Everyone whooped and cheered, clapping hands and hugging each other. All the stress of the last few days evaporated. There was no time for complacency, though. We turned straight back, to return to the previous night's camp to pick up our kit. The return journey, with the route cut and cleared, took less than half an hour.

Now we needed to get the area properly scoped out, and start coming up with a game plan. The landslide had done a lot of the work for us in getting rid of the canopy, which meant we had a decent amount of air space in which the chopper would be able to manoeuvre. There were a few big trees that needed to come down, but luckily they were dead, with no growth on them at all. This didn't translate into them being easy to fell, but Maxi got stuck in with his axe and with considerable relish, and by the next afternoon it was beginning to look as if we might have enough clearance. The next job was to take the top off the landslide in order to flatten it enough to make a helipad. This involved hacking up and moving probably a couple of tonnes of mud and stones, our machetes supplemented with hefty branches and our bare hands. We cut the branches into six-foot poles and sharpened the ends into spikes, then used these to hack up the ground and lever boulders and fallen trees aside. I'd have paid a thousand pounds for a pick and a shovel!

By the end of the second day of hacking, though, it was looking conceivable that we might well have our helipad – if the pilot was a complete gung-ho nutter. And we had no clouds. And there were blue skies at Moro, where he was based. What could possibly go wrong?

One of the essential skills of an expedition leader is being able to find a good spot for a camp. Any experienced guide will be constantly assessing the landscape around them for potential camps – that way, if you walk on for two hours and find nothing at all, you'll know where the last bomb-proof place for a camp was and be able to backtrack. Every environment has its own unique requirements and restrictions that you have to be aware of, for safety and comfort. Here in the rainforest, once you've found a potential spot with sufficient space for hammocks or tents, the first thing you have to think of is the danger of deadfall from above.

Huge rainforest trees may grow over eighty metres tall, and in other parts of the world trees can be many hundreds of years old, or in the case of similarly sized sequoias, several thousand. In the tropics, though, due to the constant climate and the rainfall that continues throughout the year, trees often live fast and die young. This means that dendochronology – the process of dating trees by counting their growth rings – usually doesn't work on rainforest trees, as they often do not have distinct growth rings. Some Amazon trees have been dated at over a thousand years old, but they are the exception. Hollowing due to death of the internal heartwood, humidity, mosses, fungus, termites, parasitic plants, huge heavy vines and a myriad of other factors mean that even the heftiest trees may be less than a hundred years old before they tumble, and even if they stay standing their mighty limbs are constantly crashing down.

Looking at the landscape around us in the crater, with its shifting watercourses, landslides and steep cliffs, it seemed highly unlikely that many of the trees would be over fifty years old. The undergrowth was littered with rotting trunks and boughs brought down before their time, often in high winds. This is something that really keeps me awake at nights. A huge rainforest branch covered with epiphytes may weigh as much as a minibus. A hundred times I've lain in my tent or hammock at night in storms and had one of those monster branches, or whole trees, crash down out of the canopy somewhere nearby. It sounds as if the sky is caving in, and is one of the scariest things in the world. To make sure you don't end up as a jungle statistic, it's essential to look skywards when searching for a site for a camp. Does that tree you're tying your hammock to have lots of recent growth and look alive? Does it have any bee, ant or wasp nests that you're going to disturb? Are there any branches far up above you that could have been eaten through by termites? Even a good-sized vine could kill you if it tumbled down from on high, so you need to be ruthless.

The next concern is flood levels. Are there signs around you that the river has risen above your camp site recently? As I'd found on my paddle trip trying to locate the Korowai, flash floods are sudden, and can be lethal. Once you've dealt with the potentially life-threatening aspects, you need to clear the forest floor, carefully removing snakes, centipedes and scorpions – nothing like as numerous or as big a problem as many histrionic adventurers and military men would have you believe – cutting back stinging and poisonous plants and making pathways that can be followed safely even at night. You make sure to use the relatively dull blade of your machete for all this work, saving the sharper blade of your pen- or lock-knife for more essential tasks. The tarpaulins go up as soon as possible to keep kit dry, and give everyone somewhere to race to if the rain gets too torrential. It's also priceless for morale to know that the camp has really begun to materialise. It's even better if someone can get to work on brewing up tea, while the others are doing the spadework. British army protocol lists putting a brew on as one of the very first things a unit should do on setting up camp. Squaddies refer to the humble cuppa as 'a cup of morale'.

Washing and water-collection points need to be assigned and made safe if you are lucky enough to have flowing clean water; if not, then you need to hang up tarps to collect rainwater. If it's a temporary camp for a small group of people, you assign a patch of forest as the toilet block. When you have to go, you take a digging implement, and matches to burn and then bury the toilet paper. If the toilet pit is going to be used for any longer than a few days, it's dug as far away from camp as logistically possible, and downstream of camp so there's no chance of faecal matter polluting your drinking-water. Hammocks are hung with thin tarps over the top, and Vaseline or bug spray is applied to the guy ropes in order to prevent ants wandering up the lines and into your sleeping bag.

Tim and I have been through this camp-building business so many times that much of it is pure routine now. I am, however, still disappointingly inept with a machete. While Maxi and Chocol will neatly slice their machetes through a thinnish tree in two clean cuts, even with my keenly sharpened British army blade I tend to hack away, turning the thing into matchsticks before it'll yield. After just a day of slashing, my palms have alarming blisters the size of fifty-pence pieces, and have to be bandaged as if I'm suffering some bizarre tropical stigmata. The younger saplings make sublime building material. They may be only as thick as my calf but they're as long as a double-decker.

In the rainforest, tree growth is all about getting as high up as possible, as quickly as possible. In a race to reach the canopy and the energy-giving sunlight, saplings grow dead straight directly upwards, with very few, if any, branches lower down. This means that an infinite supply of mast-straight timber is produced, perfect for using as support frameworks for camps. In addition, there is an endless amount of natural string, provided by stripping palm fronds and the mid-ribs of certain leaves, and you can also use the smallest vines that tangle round the trees for the same purpose.

The most treasured of jungle plants, though, is rattan. Rattan is the stuff used to make 'cane' furniture, comes in about six hundred species, and is actually a palm. But many species of rattan grow in vines, with horrid hooks to assist them in gripping on to the plants around them. The vine stems can be from a few millimetres to five or six centimetres in diameter, and many tens of metres long. To anyone who has not been hooked up on rattan, it's difficult to describe quite how evil the stuff is. Imagine getting a long line of monofilament fishing line with a pair of small fish hooks attached every foot or so, then just draping the stuff through bushes so you can't see it. It is probably the greatest cause of rainforest injuries, as it will simply never break once you get caught in it. Try and pull away, and it's you that will rip, not the rattan – as George discovered, to his distress, in basecamp when one near-tore his ear off.

This strength and flexibility, though, is of course rattan's greatest asset. You can take one long strand, cut off the hooks, then strip it at least twice or even three times, to form unbelievably flexible and resilient natural cord. The stuff is so valuable that locals will cut it wherever they find it, and carry it with them in bundles, sometimes for days on end. The flexibility and strength of rattan has other uses, though: many martial arts use it to construct their weapons, and it's also used to make 'the cane', the once much loved method of corporal punishment in British public schools.

Rattan could actually be one of the great ways to save these forests. Not only is it a wonder material which can garner much more income from intact forests than can timber, but it may also have a function in surgery. In 2010, Italian scientists discovered a way to heat rattan in a furnace with calcium and carbon to create a substance that within a matter of days can bond to bone and function as artificial bone! So far rattan bones have been grafted on to sheep bones. Initially it was just that the rattan's durability meant it was capable of taking great forces without splintering, but within weeks it seemed particles of bone started to migrate to and coalesce with the rattan. Human testing will begin within the next few years. For us in the jungle, though, rattan was just a way of tying our tarps into place and fixing the woven palm walls of our toilet.

Having only seen one place in the entire caldera so far that could possibly house a camp, I clambered up the bank on the upstream side of the landslip into the forest, expecting to find nothing more than near-vertical slime. Instead, I was rewarded with a reasonably flat piece of land that could easily accommodate us all, and would save us the hell of having to lug all our gear from the helipad to the site I'd selected earlier. It is impossible to overstate the good fortune this represented. That one piece of ludicrous luck turned a punishing and quite possibly unsuccessful mission into an achievable goal. Taking care to keep the largest trees intact, we got to work on clearing a decent space, so we'd be able to get straight to work as soon as supplies arrived on the helicopter.

16

Through the gorge ahead of me rushed a jade river. Black walls overhung the water, presenting a formidable gateway to the land beyond. No local would ever attempt to swim up through that furious torrent. What lay beyond would be unseen by human eyes. Taking a deep breath, I ducked beneath the water and swam hard against the current. When it thwarted my progress, I tugged myself along the bottom using boulders as handholds, then burst to the surface. There, amongst the tumult, I struck upstream, pulling myself up on to a rock. I was now in another secret gully – yet another place I could state with confidence no one but me had ever seen, a place of luxuriant beauty.To one side of the river, mossy fallen trees lay in a tumble like so many matchsticks. On the other, great boulders the size of small houses. When I looked up – the walls of Bosavi, high, high above me. I was transfixed by the clouds and the mist billowing, retreating, seething amongst the rocks and the tree-lined ridges. It was truly breathtaking – after a moment or so, I was so entranced I had to remind myself to breathe.

Bosavi's near-perfect caldera has kept its fauna trapped inside for at least two hundred thousand years – back then was approximately when it last erupted. Its altitude means the animals within will be montane species, and therefore different from the lowland species in the surrounding land. In evolutionary terms Bosavi is an island, and a hotbed for speciation – that is, the development of new species from older lineages.

Our mission in Bosavi, as I said earlier, was a simple yet grand one: to find creatures that'd never before been seen by science.

After a few days finishing off the camp and making it habitable, Gordon arrived in a helicopter with medic Jane and scientist Chris Helgin. Chris is a noted mammologist from the Smithsonian Institution and a global expert on New Guinea's warm-blooded fauna. He is also the protégé of my idol Tim Flannery, and one of only two or three people in the world familiar enough with the wildlife of New Guinea to be able to look at an animal and tell almost straight away if it is a new species. Immediately, Chris took me by surprise. I was expecting a man of such prestige to be in his fifties or older, but instead he was probably a bit younger than me, which was slightly sobering. He didn't look like a jungle man at all and clad in welly boots, long-sleeved brown T-shirt and slacks. Initially, he looked a bit out of place.

There was, though, no doubting his credentials, or that this was a man who really knew what he was doing. But there was immediate concern when he arrived sick with fever and headed straight to his bunk. He lay there for two days, and we were beginning to talk about evacuating him. It must have been unspeakably miserable for him, no more than a few days out of the US and he found himself stranded in this rough camp in the land that time forgot, unwell, unknown to the rest of us and as yet unaware as to whether he was even amongst friends. It must have taken all his reserves not to have got straight back in the chopper and headed home.

As assistant, Chris had brought with him the wonderful, cheerful Musee, who we'd all learnt to trust implicitly and was our perpetual saviour at basecamp. Cameraman Robin Smith, Nick the soundie and production team Steve Greenwood and Jonny Keeling completed the team.

In the helicopter storage hatches were our tarps, the portable generator and – Hallelujah! – the boxes of food. But our initial elation turned to outrage when we opened the boxes to find that Tony the miserable Australian chef had kept back all the decent food for his contingent of the expedition, and just given us bully beef, mackerel and rice, and not enough of that. On the second helicopter flight in was Chief Seguro from Sienna Falls, and on the third were elders from Iggycelabo village outside Bosavi to the south. Iggycelabo is no more than seven or eight miles away as the crow flies, yet it's five days' hard walk away. They hadn't seen their relatives from Sienna Falls, to the north of the crater, in many years, and greeted each other in the *wontok* language of Kosowan, spoken by less than five hundred people (and a true language, not a dialect). In this respect New Guinea is without doubt the richest place in the world, with more languages than all of Europe, Africa and the Americas put together. Both Iggycelabo and

Sienna Falls villages have no more than ten adult men – brothers, uncles, cousins. When they meet, they brush noses with each other like Eskimos, while making gentle grunting noises – as tender a greeting as I've seen anywhere in the world. There was no mistaking the genuine joy they felt at being reunited with old family and friends.

The next ten days were spent exploring the surrounding forests, laying traps and using old-fashioned legwork to try and record as much as possible about what was living in the area around our camp. First signs were freakishly positive. The first night out, we spotlighted three cuscus in our spirit tree right in front of camp. They weren't doing much, as they were kind of transfixed by the torchlight, but in one single tree we had equalled the mammal count of three weeks spent at main basecamp outside of the crater!

The very next day, we were standing on the helipad in the sunshine when Jane started tentatively calling: 'Errr ... is that a ... no ... err, guys, it's a ... tree kangaroo ... it's a tree kangaroo!' We walked towards her, expecting to see a distant silhouette way up in the canopy, but there it was, wandering up the creek and heading straight into camp, as nonchalant as any animal you'll ever see. It was a Doria's tree kangaroo, dark grey with a russet chest and a silvery tail, its posture and shoulders bringing to mind a little bear, but its lazy hopping gait unmistakably marsupial. We looked around frantically – the thing was no more than ten metres away from us, and we had to get it on camera. But there were no cameras to hand, and everyone over the creek in camp was ensconced in the trees and utterly oblivious to what was happening. We couldn't run across in front of the creature because we'd have freaked it off into the trees.

I started loudly stage-whispering – 'Luke, Luke ... *Luke! Luke!*' – as the kangaroo made to hop right into camp. And then, two of the local guys noticed him and all hell broke loose. They started whooping and screaming, and stampeded down the bank, corralling the startled animal. Then as it turned to run, they grabbed it by the tail. '*No!*' we all yelled simultaneously, seeing our treasured subject about to end up as lunch.

I slid down the bank, yelling at the local guys to stop, pushed them aside and grabbed the kangaroo off them by the tail, just as Luke appeared from the trees with his camera. This was not the scenario we had in mind. Here was I, sat in a stream, holding a bewildered kangaroo by the tail as it scrabbled about, clearly having realised it was in a spot of bother.

The animal was obviously distressed, so I let it loose. It sprinted off in muscular fashion, spinning like the wheel of a rally car in the dirt, but then within a few metres it stopped to figure out exactly what had just happened. It must have been a bizarre moment: he'd never seen people before, or in fact anything that posed a threat to him. Then he was suddenly grabbed, then released, and it must have been totally bewildering. As he loped away we got some great footage of him, but somehow the whole thing left us all with the feeling that we'd ended up frightening an utterly wonderful and extraordinarily rare animal.

That night in camp we sat down with all the locals, and reiterated our rules. The animals were not to be manhandled, and certainly not killed. But it was clear from the blank, uncomprehending faces in front of us that to them this was total madness. Luckily, though, it was not our last encounter with the Doria's tree kangaroo. Just the next day, a shout went up from the slopes above camp, and there was Gordon, mobilised, with his powerful lens and a tripod. He'd managed to find a spot almost level with a kangaroo in a nearby tree, and though the animal was rather shy, going out of its way to put foliage between itself and Gordon's lens, the footage he got of this creature grooming, feeding and moving so easily through the branches was heart-warming and bizarre in equal measure.

Doria's tree kangaroo is the largest marsupial found in New Guinea's central mountains, and as such is particularly susceptible to over-hunting. It was a really good sign to have found two of them within our first few days. I was probably quite lucky to have got away without any injury. Maybe it was because the specimen I'd encountered was in such a state of shock. The guidebooks tell tales of Doria's tree kangaroos eviscerating hunting dogs and even hunters themselves who have tried to catch them. They are extremely odd creatures, very clearly having evolved through most of their history to be terrestrial, but then for some unknown reason taking to the tree-tops. Perhaps it's because they moved into environments where – as in the crater – there was little open ground to graze upon. Alfred Russel Wallace comments upon their peculiar evolution in his book *The Malay Archipelago*: 'The leaping power of the muscular tail is lost, and powerful claws have been acquired to assist in climbing, but in other respects the animal seems better adapted to walk on *terra firma*. This imperfect adaptation may be due to the fact of there being no carnivora in New Guinea, and no animals of any kind from which these animals have to escape by rapid climbing.'

They look awkward, both on the ground – where their heavy tails seem to tip them forward so their noses are almost scraping the dirt – and in the trees. They hop about on their large rear limbs, and grip around branches in koala-like fashion with their smaller forelimbs. The tail is not prehensile, but is long and heavy for balance, the toes have rubbery soles and long sharp claws for gripping, and they have heavy bones to enable them to leap down perhaps three storeys' height from the trees without injuring themselves.

Our mammal encounters got better and better by the day. First off, Jonny Keeling found part of a marsupial skull and had the presence of mind to bring it back into camp. Hours after we gave it to Chris, he was poring over books and measuring the teeth and upper jaw, with growing excitement.

'So where did you say you found this, Jonny?' Chris asks, sat in front of a laptop with scores of skull pictures registering on the screen.

'It was just a few hundreds of metres upstream, I was having a swim and found it alongside the pool. Didn't think much of it really, I mean it's only the top half of the skull – can you tell anything from that?'

'Well, yeah! Most of what we do is from the teeth – that's probably the most likely bit to be diagnostic.'

By 'diagnostic' Chris means that this is the part of the animal that is most likely to prove unique to a particular species. Often the dentition, or arrangement of the teeth, and skull shape and size are the best signs as to what a creature actually is. It's often far more reliable than fur colour and size, which can be vastly different in individuals of the same species.

Chris took the fragile piece of bone in his hands. 'It doesn't fit,' he said. 'It's quite similar to a silky cuscus skull, but it's missing premolars here and here' – pointing to a flat gap in the jawbone. 'I reckon this animal is new.'

My response to that on camera was to enthuse: 'This is one of the most amazing discoveries we've made, we just need to find the animal this skull belongs to.' Then I pondered the implications of what I'd just said for a second: 'Well not that exact animal – that one's dead.'

The day before, Gordon had filmed another arboreal mammal, a cuscus, emerging from a hollow tree trunk. The footage was now examined again with new interest.

'Look at this,' Chris said, 'this dorsal stripe here doesn't continue down the back, and the fur is definitely almost black ... I mean, obviously that doesn't prove anything yet, but it *could* be our mystery animal.'

With a subject and a purpose in mind, we mobilised everyone in camp: the local guys headed off enthusiastically into the bush, and everyone else set off with cameras at the ready. Hands were thrust into holes in dead trees, just to see whether anything interesting inside would bite you. A hundred humane traps were set, using up our entire remaining supply of peanut butter as bait, and every inch of hillside that wasn't too steep was tramped over by bare feet and wellington boots.

Bosavi had a totally different character over those first few sunny days. When Tim, Luke and I had battled our way into the crater, it had felt like a brutal, unforgiving place that didn't want to be explored. The ceaseless rain, the steep terrain and the evil plants had battered our spirits in just four days, yet here we were, in glorious sunshine, thinking ourselves wanderers in a magical Eden, unspoilt, unexplored, untamed. My first aim was to try and map the course of the main river, heading both up and downstream from camp, slipping and sliding over the rocks, wading and sometimes having to swim through the torrent in order to proceed.

Progress was slow, but fabulous, all the more so as I knew that the local people would not have attempted to swim some of the faster-flowing sections, so I knew in my heart we were the first ever to see them. On either side vertical rock walls forced us into the river, and the volume of water running through it meant we would have to zigzag backwards and forwards across the flow continually. With over twelve metres of rain tumbling into Bosavi every year, it has some of the highest rainfall of anywhere on the planet, and all of that rainwater filters down through the volcanic rock and pours straight out into these glimmering creeks.

We were wading in water purer than Evian, swimming through the malachite and jade pools, treading cautiously over mossy fallen tree trunks that spanned the waters. Yet again, the vegetation was utterly prehistoric, and it was impossible to shake the feeling that you were wading through a film set. Generally speaking we were closeted by the walls of the canyon, but every once in a while I'd turn a corner, and towering over me would be the crater walls, well over a thousand metres above. It was impossible to take your eyes off them. It was like finding you'd been walking under the north face of the Eiger all day, but not noticed till now. It was totally hypnotic, watching the clouds and the mist up there amongst the rocks.

Gordon set up his gear on the riverbank to film a family of torrent flycatchers, striking white birds with black trim that zip backwards and forwards above the rushing water, catching insects on the wing. One bird, with slightly fluffier, less defined plumage showing flecks of brown, was obviously a fledgling, being tutored by his parents.

The most dazzling of daytime spectacles, though, were the butterflies. Small blues were the most numerous. Should you have left a sweaty sock in the sun to dry, when you returned it would be studded with blues, like scales down the back of a basking iguana. If anyone wandered past, they'd take off simultaneously and circle about in a shimmering cloud for a minute or so, before settling back down to lap up the precious salts. Most impressive were the birdwing butterflies, aptly named, as they are easily a match in size for a small bird. There are flouncy ladies' lace hankies in white with black trim, super-sized fritillaries with intricate black veins on a shining copper background. Most spectacular of all the butterflies, though, is the neon-blue birdwing, of a colour more intense than any other in nature. The blue morpho butterflies of Central America are some of the most prized in the world, but would look dull by comparison. I cannot take my eyes off them. However, as soon as they perch, the black underwing folds up and they disappear. They are a miracle.

It was the end of the first week before we worked up to doing a full-on night walk. First we needed to get more of a feel for the land around camp before we could go out at night safely, and also to clear some paths so we could go anywhere we liked. We all knew that it would be the night walks that gave us our finest bounty.

No sooner had we crossed the helipad on our first attempt than Nick the soundie called out: 'Eyeshine, what's that?' Right in front of me, at eye height, was a coppery ringtail, a species of marsupial that spends its nights scouring the boughs and branches for leaves and fruit. It was dazzled by the torchlight and just sat there dumbstruck, looking a bit like a fat hamster who'd got stuck in a tree and now couldn't get down. The same night, Gordon found a painted ringtail, which is similar in size and shape but has glorious golden and alabaster-coloured fur. There was already no doubt that Bosavi was busier with warm-blooded life than anywhere else any of us had seen in New Guinea.

As the man in charge of editing this whole expedition into three hours of television, Steve Greenwood was eager that the viewer should get a proper view of the hardships we were going through. Having filmed all the leeches we could possibly shoot and a venomous scolopendra centipede inside my hammock, Steve mused: 'What we really need is some rain, to make it all look like we're suffering.'

Everyone looks to the heavens – why can't he just be content with us enjoying ourselves in paradise?! That evening Maxi comes into camp as we're sitting down to eat our dogfood dinner, and points to the sky: 'Wind 'ee come, big pla baggerup,' he tells us. Just minutes later, wind does indeed come, the near end of a storm front that brings rain in biblical amounts. The deluge is so intense that we have to yell to be heard above the roar. Wandering out beyond the shelter is like being hit by a pressure hose, soaking you to the shivering skin instantaneously. Water pools by the dustbinload in bulges on the tarpaulins, and we have to busy ourselves scampering around and pushing the water out – otherwise, the whole camp would collapse around our ears.

After the initial cloudburst of half an hour or so, the rain eases to become less forceful, but persistent all the same. For that afternoon we take shelter under the tarps and sit around chatting or writing up our diaries. But soon it becomes evident that this is not going to be merely a spell of drizzle, and that we're just going to have to get back out and do our jobs, even though that'll mean being ever after soaked and miserable.

The rain hammers down non-stop for most of the remainder of our time in Bosavi, making the frogs and leeches very happy but chipping away at our spirits, giving no opportunity to dry out our clothes or ourselves.

In scale, Camp Bosavi is nothing like our main and original basecamp. We have one blue tarp about ten metres long to serve as a kit tent, and another about the same to function as a kitchen and eatery. The dormitory is slightly larger, but not large enough to escape the thunderous snoring of certain members of the crew (to remain unnamed).

In basecamp proper we elevated ourselves above the mud with wooden boardwalks, but here we have to stomp over the same paths time and time again, and soon after the rains start, the mud becomes truly impressive. We place train-tracks of sticks into the squelch to give us a little grip, but still wallow around everywhere in ankle-deep ooze. The crew take to their welly boots, but I unfortunately have forgotten mine. I get another pair sent in on the helicopter with Gordon, but with my bad ankle refusing to straighten I can't fit into them. In an attempt to make them serviceable I take my knife to cut off the upper sections, to help me get my feet in. In the back of my head I tell myself: 'Don't cut too low, you can always cut off more later. Just whatever you do, don't cut too low. Anything but too low.' Then promptly slice the rubber so severely that they end up as green slingbacks, slopping about ludicrously.

As we all aim to keep a pair of dry trousers for the evening, in the mornings I tiptoe around the mud in lycra cycling shorts, a thick fleece, and these comedy shoes. While preparing lunch in soot-blackened pots, I inadvertently smear the culinary eyeliner above my cheeks. I look like an extra from *Lord of the Flies*. As mud and carelessness set in, though, it's not just me that slips into camp fashion crimes. Jonny Keeling is often seen wandering round in just a pair of James Bond speedos and flip-flops. Having given his head a grade-one shave all over, he looks a bit like a commando heading out for a mud wrestle. Nick strolls about camp with a roll-up ciggie hanging from his lip, wearing garish surf shorts and floppy wellies, his head torch set to one side of his head so it doesn't dazzle people as he talks to them. He looks like the village simpleton. This look is finely set off when he's recording sound in the rain. A camouflaged poncho covers his sound kit, giving him the appearance of a military Teletubby.

Bossman Steve Greenwood's glasses have got slightly bent in the middle and sit askew on his nose. As his jungle shirt and trousers get more and more waterlogged, they hang off him, so heavy he seems to struggle to lift his limbs. He calls to mind a bedraggled, gangly schoolkid who's had to borrow his older brother's kit to play rugby in the rain. None of us care, though. In basecamp one of the female scientists is given to scooting off and putting on make-up and doing her hair if she thinks she might appear on camera. I, for my part, go through two days of filming unaware I have a huge white spot on the end of my nose and small chunks of bully beef nestling in my beard.

Small injuries were also starting to get under people's skin – literally. Tiny burrowing mites called chiggers drill beneath the skin around the ankles and under the waistband of your trousers. It itches like fury, and occasionally the lesions rub and rub till they start to go yellow and septic. Old leech bites and rotting trench foot also itch, not furiously but continuously. It's not enough to produce a tantrum or to send you in search of hardcore antihistamines, but it's enough to gently, slowly, drive you to the margins of insanity.

After the first week everyone in camp picks up an infestation of intestinal worms. In all probability this comes from the locals (who have the worms as a matter of course) handling foodstuffs with hands that haven't been washed after going to the toilet. They make you feel really run down, and require a course of large gut-rotting tablets to get rid of them. I manage to get my first embarrassing injury while shinning up a tree to fix one of the camp tarpaulins, severely chafing myself in what Luke calls 'the honeymoon suite'. It doesn't look like much to begin with (the wound, that is, not the other), but as with all open wounds here, it starts to fester and I end up before going out for walks having to apply wrappings of sticky tape around a part of the body you never want to bandage. Or remove sticky tape from.

Our curious habit of expending enormous amounts of effort on finding animals which we then catch but don't eat puzzles the clan chiefs. There are four of them, Seguro and Dixon from Sienna Falls, Caseo and Caliwayo from Iggywacu, all in their late fifties and with unusually expressive faces. They wear old shorts that look as if they've never been washed, crazy headgear ranging from an old leather patchwork Worzel Gummidge hat to a knitted sock, and extravagant beards. In the swirling smoke they look for all the world like members of an old Deep South blues band – Caseo the Lips and Howling Dixon Jefferson. They all have pierced septums, through which they'd wear bones or carved shells at a singsing or birthing ceremony. Halfway through our jaunt, after several days of non-stop rain, we sit around the campfire and chat to them about their early lives and their first contact with the outside world. Chocol translates for us in his elaborate slightly nonsensical English, which makes their answers even more peculiar. In a culture that has no books, storytelling is the main source of entertainment, and stories are told as if the narrator is reliving the tale, throwing themselves around, re-enacting sound effects and miming unrestrainedly.

'So what was your life like before the white men came?' I ask.

Chief Caseo begins his story: 'Before the white men, we were always moving, hunting, but then the missionaries came and taught us to live in villages. We came to Bosavi sometimes when we have a big singsing to catch many animals for eating.'

'Do you remember what animals you used to see here?'

The cassowary, they recall, and the echidna, and then they remember their old friend the tree kangaroo and leap into a tableau depicting a kangaroo hunt. Dixon pulls an imaginary bow back, and Caliwayo jumps into a spindly tree nearby, clinging to it with the exact posture of a tree kangaroo and mewing down at the hunter beneath him. It's so strange to see two old men in positions of power performing this bit of physical theatre, and has clearly been honed over the years spent telling hunting tales round the fire.

'And do you remember the first time you saw white men?'

The elders nod enthusiastically – of course, yes, they were just young boys back then.

'And how on earth did everyone react to seeing them?' I ask.

Their response comes not in the form of words, but in another dramatic mime. The two old men run off into the corner of the tent together and huddle in the foetal position with their arms round each other, cowering and terrified, then show how their fathers had spirited them off into the forest. Seeing a white-haired chieftain doing a somewhat over-the-top but totally serious portrayal of a shaking, wide-eyed baby, was totally at odds with my preconception of the reticence and dignified reserve I expected from an elder.

'And were your fathers warriors?'

'My father was the greatest warrior,' Seguro states boldly, striking his chest. 'He killed many hundreds of men.'

Caseo continues: 'They were always at war with other villages, big wars . . . and sometimes they would eat human flesh!' This was unexpected – well, not unexpected that they would have been cannibals, but unexpected that they would so readily volunteer the information. I was planning to talk with them for many hours, building up their confidence, before I risked broaching that subject.

'I remember our fathers taking the man' – a seated Caseo pulls his legs up to his chest – 'and cutting here and here' – he makes imaginary score marks down the front of his thighs. 'They took hot rocks from the fire and cook them in mumu.'

The mumu or earth oven is usually a pit filled with embers or heated rocks, then green vegetation, then meat, then more vegetation, then water, then filled in with earth. In a few hours you have a well baked meal.

'And was that so you could take on their strength or their spirit?' I ask.

Caseo looks at me in a confused manner and asks for an explanation from Chocol, who seems equally nonplussed.

'It is for food, the taste is very good,' he replies, as if this is the most obvious thing in the world. With a little more gentle questioning it becomes clear that in this corner of New Guinea at least, cannibalism does not seem to have the macabre spiritual associations it has in other parts of the world. Instead, it was seen as merely a valuable source of protein in a place that has relatively little natural meat.

In the mid-twentieth century, however, the gruesome practice of cannibalism for food resulted in a horrifying epidemic amongst the Fore tribe of the Eastern Highlands. The Fore would actually consume the dead of their own tribe, in order first to return the life force of the deceased to the community, but secondly to get hold of vital meat. Visiting anthropologists became aware of the epidemic sweeping through the Fore. They called it *kuru*, which means 'to shake', although they also called it the laughing sickness, due to bouts of uncontrolled laughter the afflicted would suffer, along with blinding headaches.

Kuru was incurable, and as outsiders started to analyse it they realised it was a transmissible prion disease, a human spongiform encephalopathy in the same mould as CJD or Creutzfeldt Jakob Disease. Much as it was suggested that cows had become infected with 'mad cow disease' by eating infected ground bonemeal from other cows, it seemed that the Fore people had contracted *kuru* by eating their dead. While the men of the Fore tribe had eaten the best cuts of the meat, the women and children who washed the corpses and consumed the leftovers, including the brain, were eight times more likely to be affected. In addition, *kuru* victims themselves were particularly prized as meat, because those that had died quickly were left with a thick layer of fat, and were described as tasting agreeably like pork. With an incubation period of five to twenty years, the implications for the Fore were pretty horrifying, but as Christian missionaries spread throughout the region quelling cannibalism, *kuru* also started to decline.

While we may not have slipped to the level of eating each other, camp cuisine was beginning to take its toll on us all. The vegetables we brought in had gone off, the peanut butter had been used as rat bait, and thanks to the locals' habit of putting ten sugars in every cup of coffee and eating five packs of crackers for every lunch, the rice, bully beef and tinned mackerel were all that was left. The camp was split neatly down the middle between those who were made physically sick by the thought and smell of the fish, and those who would rather die than eat the beef. I fell into the latter half. Every evening there was an odious hour when the subject of food came up, and fantasies and fetishes peppered the disgusted exchanges about what the next meal would contain.

'So what will Sir be choosing off the menu?' inquired Steve Greenwood in the manner of a Maître d.

'Well, it's a toss-up between Bully Beef Fricassée presented on a bouquet of rice, or sardines au naturel . . . or you could really mix it up and have the rice on top.'

'So it's dog food or cat food, basically?' Having asked this question in jest, the image was then fixed in my mind and the fare on offer seemed even more unpleasant.

'God, you wouldn't serve bully beef to your dog, you'd have the RSPCA round for animal cruelty,' said Gordon.

'What do you think's in that stuff, anyhow?'

'Mostly udders,' Gordon surmised, looking into the fat-smeared can.

'I think there's a fair amount of hair and unmentionables in there too, mate.'

'Lips and arseholes,' said Luke, with typical South African reserve.

'God, mate, I'd rather eat that than the sago!'

'Sago?' asked Luke, the only one of us lucky enough not to have had to eat it.

'It's pulp from the inside of a tree, soaked in water, squeezed and dried – tastes kinda like crushed blackboard chalk,' I explained.

'Or like wallpaper paste, only not so tasty.'

'We had it fried, it was like chewing on a car tyre,' new cameraman Robin contributed.

'And what about the sago grubs?' These are beetle larvae the size of a wriggling kiwi fruit that live inside the sago palm and are eaten raw as a delicacy. 'Have you had those?' inquired Luke.

'I have,' I admitted. 'Cooked, they're pretty tasty, kind of like scrambled eggs. Alive isn't so good, kind of like raw egg in a chewy skin, and the head is just nasty and crunchy.'

Yucks from everyone.

'Tell you what, though, what I need is some fruit – or a salad. How long do you have to go without before you get scurvy?'

'I miss the booze most, man.' Nick the soundie had admitted some weeks earlier that he was 'already digging deep, man', and that was *before* the tiny supplies of Scotch ran dry.

Now he was really pining. 'Nice bottle of Cabernet Sauvignon and a huge chunk of rare steak – I'm talking blue, man.'

He's not lying. On a recent trip to Lousiana with me he sent back a steak that was so rare it had barely stopped mooing, saying it was overcooked. The chef all but chased us out of the restaurant with a carving knife.

'Just a huge chunk of steak – touch, touch,' Nick mimes, touching either side of the fantasy steak, for a sizzling second on a hotplate. 'Dunk it in some horseradish ... *num.*' He scoffed the imaginary steak down, in both hands like a giant sandwich, before drawing his sleeve across his face like a Neanderthal wiping blood from his chin.

'It's a beer for me,' Robin mused, 'icy cold so the sweat's trickling down the outside of the glass.'

Everyone groaned.

'We *must* have something – has anyone ever tried snorting coffee?' Nick again. Joking, I assumed.

'Jane's got adrenalin in her med bag, I've seen it,' Gordie chimed in.

Jane looked silently but warningly at him.

'Chris has said I can have some of his preserving alcohol,' said Nick, 'but only if I drink it out of the bottle that's already got a dead specimen in it.'

More yucks all round.

'Come on, it's not that bad, it'd be just like vodka if you watered it down a bit.'

'That's what winos say about meths.'

Nick then brought up a food anecdote from the Borneo expedition. 'We'd been going for two weeks up this river, eating rice and noodles and nothing else, and every night all we were talking about was food. The local guys were killing all the animals before we had a chance to film them, it was a total nightmare. One day this guy comes back with a mouse deer—'

'The highly endangered, cute and cuddly mouse deer,' I interjected.

'Yeah, but man, it was delicious.'

I take on the overhyped tones of a movie-trailer voiceover. 'The BBC is leading an expedition into the Heart of Darkness. These animals know no fear of man, they have never been hunted . . . until now.'

'Yeah, it does sometimes seem we're just on a gourmet wildlife tour of Borneo. But thing is, right, they smoked up this mouse deer and gave us a big chunk each, and it was so weird, man, everyone had been eating together, but the second we had meat, we all took it away to a little corner on our own.' He mimed a caveman with heavy brow carrying off his meat, turning his back to protect it and throwing suspicious glances over his shoulder. 'And nam, nam, nam, we'd scoff down our meat, and the *second* we'd finished eating, we stopped talking about food, and started talking about sex!'

As we hadn't had any meat and there was no prospect of any, the conversation continued for another hour or so and every foodstuff on the planet was covered, before we got on to clean sheets, hot baths – and then came the more sombre comments.

Jonny Keeling got out a picture of his wife and their beautiful baby girl, and went very quiet and contemplative. It was almost uncomfortable seeing a guy who looks like he should be in the SAS getting openly emotional in front of his mates. 'I think I'm doing fine,' he said, 'then I look at these photos and it's like someone's just taken my heart and given it a big squeeze. It's really hard.'

Gordon had a picture of his little girl as the screensaver on his computer, which kept flicking up while he was checking the footage from the camera traps. His daughter has huge, beautiful eyes and stares back mournfully against a sunset on a Scottish beach. She summed up the perfect image of home.

The crew are all, to a man, physically striking guys, with equally attractive partners, so it's not surprising that their children are all so stunning. That they're all professional cameramen also helps explain why their holiday snaps look like Calvin Klein posters.

'I can't do it, man,' Nick admitted. 'I can't even look at my photos. It just messes me up and brings me down. It's better if while I'm away I just try and forget home. Any time I start thinking about them I just want to say 'stuff this' and get on the next flight home.'

'You mean the helicopter, boat ride, walk, helicopter and then five flights home. Which we can't do for fifteen days.'

'Fourteen days and nine hours by my watch,' says a miserable Nick. But I understand completely. There's something about the images, perhaps the fact that they're taken on snowy, windy days back home, or down the local park or making sandcastles on the beach. They're so very different from everything that surrounded us that they seemed to be from another world. One a million miles away. Dinner tonight was quieter than usual. There were no quips about the bully beef, everyone was off in their own thoughts, counting the days till they would be home.

Nick is one of my closest friends, and we've spent many months on the road together. I love travelling with him, as he's always ready with a quip and a joke, or a sarky comment that never allows anyone to take themselves too seriously. He is the only Englishman I know who can use the words 'man' and 'dude' in the same sentence without any sense of irony whatsoever. None of us had ever seen him this down before. He was utterly hating the jungle, and his negativity was getting infectious.

'I literally haven't seen the sun for a month, man,' he said, scraping his hair upwards in a slightly manic manner and rubbing his stubbly chin. 'And my beard's coming through all white, look, this place is making me old. I look like some wino sat outside Budgens scavenging dog ends. I hate the jungle, man. This is the last time I'm ever coming to the jungle. Expedition Maldives, that's me. Expedition Maldives searching for a new species of fish that lives off an island where the only place to stay is a five-star hotel. Otherwise you can get stuffed, man.'

On one of my solo recces upriver from camp, I spotted what looked like a cave above the waterline. It was tough getting upstream with all the cameras, but the only way to move through the terrain was by wading through the river. The rocks were as slippery as if they'd been greased up with motor oil, and one little slip could have dunked the camera or sound kit, leaving us in deep trouble. In order to penetrate far enough, we needed to swim through narrow gorges with all the kit wrapped up tight in dry bags.

When we finally reached the site below the cave, it became clear that the only way to climb up was via a short but sheer rock face, with the only handholds being thin plant roots. Worse, when we actually reached the cave it was surrounded by single-stemmed plants with huge heart-shaped leaves. These were stinging trees, the most evil plant in the world. These living nightmares have hairs much like those of stinging nettles back home, except that each hair is a hollow silica hypodermic filled with neurotoxic poison. The sting at the time is no worse than a nettle, but every time you brush the stung area, or sweat or get it wet, it starts burning again – for three months. I am not joking. It's the gift that just keeps on giving. You can get rid of some of the effects by applying leg wax to the affected area and ripping it off, but none of the crew were armed with leg wax at that particular moment.

The cave itself was, well, not actually much of a cave. It was more of a hollow, and quite a soggy one too, so not much of a shelter for animals. However, on one wall there was something pretty extraordinary: two rather beautiful moths sat there motionless – motionless because they were dead. Bursting from their bodies were some curly, tangled, wispy strands – rather like miniature grape vines. This was the cordyceps fungus, one of the most grotesque manifestations of parasitism in the natural world. The fungal spores are ingested by insects, and then begin to grow within their hosts.

As a result, the insect's behaviour becomes erratic. In some ants, the cordyceps will affect them in such a way that they climb to a high point, which favours the next step in the fungus's development. It erupts from the exoskeleton of the insect, then spores burst out of the fruiting bodies to infect other insects. Cordyceps are prized in Chinese medicine: in fact several athletes who won golds at Beijing were said to be fuelled by cordyceps fungus, which can cost over $1,000 an ounce. Given its highly macabre life-cycle, this seems to be quite a leap of faith. I know they only infect their own specific insect host, and that I personally am perfectly safe from them, but all the same the thought of a vile fungus growing inside me is the kind of thing that'll wake me screaming in the middle of the night.

As our mammal search intensified, images started returning from remote cameras that we had set up around the likely ridges, game trails and tree hollows. They displayed a weird and wonderful mammal fauna. There were antechinus, small, hopping carnivorous shrew-like marsupials also known as pouched rats, which are most notable for the male's habit of copulating to the death. Females may live for two or three years but males only manage one, living through to maturity then mating continuously for a whole day, using up all their protein and fat stores in one mammoth copulation session which invariably kills them.

Perhaps most tantalising, though, was a huge shape with burning eyes, bigger than the tiny marsupials we had encountered but with an unmistakable rodent-like demeanour. It may sound weird, but one of the most impressive creatures that inhabit New Guinean forests are the rats, and in particular the giant woolly rats. It's all down to the curious evolution of the species here in New Guinea. There are no primates here, no big cats, though there are feral dogs in some places, including the famous 'singing dogs' that howl through the mountains of central New Guinea. The marsupials and rodents have squeezed themselves into the niches left behind by absent animals. So you get kangaroos that scramble round the tree-tops like monkeys, and rats that stalk the forest floor like badgers or foxes.

All of a sudden, while looking through the shots from one of the camera traps, Chris started getting excited: 'Wow, look at this! The long naked tail – what do you think that is?'

Gordon leant in closer. 'It just looks like an enormous rat.'

'That's exactly what it is, a giant woolly rat.'

Gordon did a double take: 'Jeez, it must be about this size' – he held up his hands to indicate an animal with a body the size of a Highland terrier.

Chris nodded. 'And with the tail it'd be over a metre long.'

Gordon paused and chuckled, then looked at Chris and realised he was serious.

Chris continued: 'It would have to be the largest rat in the world, but this is a mountain species. You won't find this on any of the low areas round here – it would almost certainly be unique to Bosavi.'

'This image isn't going to cut it, though?'

'No, we'd need to actually find one in the flesh. But there's no question in my mind. This is a giant woolly rat, and I suspect Bosavi has its own kind.'

Everyone stopped for a second as the resonance of that last comment sank in. To find the largest rat in the world inside an extinct volcano never before explored – it sounded like something out of a Michael Crichton novel.

After that, our efforts to capture mammals redoubled. We headed out on interminable walks every night, tramping, slipping and sliding through the mud, our huge torches trained on the ground rather than up into the tree-tops, scouring the leaf litter for our rodent superstar. Not surprisingly for a group of alpha males who all prided themselves on their wildlife skills, we formed into two quietly competitive camps. Gordon and Jonny Keeling headed one team, and Steve Greenwood and I took charge of the other. We went off separately with a camera in tow, putting in more and more hours in the field as the days passed, getting further and further from camp, each of us frantic to be the crew that would find the first of our unknown animals. Whenever the other team returned triumphant, we publicly cheered them while quietly cursing that it hadn't been us, and vowing to get straight back out and put in even longer hours.

As my team scaled a small waterfall one night, the spray bounced up in the torchlight revealing huge spiders' webs, glistening with diamond-dust drops. The ruby eyeshine of moths and the greeny glint of spiders' eyes told of facinating critters to be found and identified in the darkness. The biggest of these spiders was also the biggest I'd seen in all my many months in New Guinea, a rival to the giant bird-eating tarantulas of South America. It was a trapdoor spider, its abdomen the size and shape of a slightly squeezed ping-pong ball. Its cephalothorax (head and thorax combined, as found in most arachnids) was like a fifty-pence piece. Both its backside and legs were a metallic gunmetal blue-grey. As I went to pick him up, he reared up on his back legs revealing fangs like cats' claws emerging from a blood-red maw. He was the stuff of an arachnophobe's worst imaginings, but for me the stuff of dreams.

As I was chatting with the guys about this magnificent creature, suddenly a squeaking and clicking started upslope, then a rustle, as a shape under the leaf litter moved towards us. 'It's a mouse,' said Luke the cameraman, just as it crawled out of the darkness while I sat there intrigued. But it wasn't a mouse at all. Dragging itself along on its elbows, and wings tucked in close to its sides, was a tiny bat, scrabbling along the ground like a snuffling bloodhound with its nose stuck in the leaf litter. While we knelt there speechless, the somewhat grotesque creature crawled up my leg, across my lap, down the other side, then carried on scrabbling. I was for a second stuck for any words or interpretation. Was it stunned by our lights? Had it come out of a low roost and was it looking for a way to get airborne? But then, as I pondered such things to the camera, another bat scrabbled out of the mulch, and then another one!

Three bats on the ground, evidently echolocating, the sound too high pitched for us to hear; but they were also squeaking stridently, and all of them moving with purpose. I reached down, and grabbed one behind the shoulder blades, pinning him with his head forward. Good job too, as he opened a lipless slash of a mouth to reveal vicious teeth that would easily have bitten me to the bone. Bats in this group are called mastiff bats, named after the fighting mastiffs that bite and never let go. My first instinct was to give him a helping hand, placing him on a low tree and allowing him to fly off, which he duly did.

Later, back at the squelchy mud fest we called camp, Jonny Keeling brought up an interesting possibility. He reminded us of New Zealand's lesser short-tailed bat, which has evolved without any ground-based predators and finds an endless supply of terrestrial insects. They do a good deal of their hunting scrabbling about on the ground. They use their wings as forelegs and have developed thumb and toe talons for a better grip, and will even excavate burrows on the ground to roost in. Vampire bats are the only other bats that have the peculiar four-legged motion on the ground; they land near their prey, then scuttle over to get stuck into a hearty blood meal. It's a behaviour unknown anywhere else in the world, but here we'd found three bats, on the ground, able to get airborne if they chose to but staying down low, echolocating for insects as they moved. Could this be the first recorded instance of bats hunting insects terrestrially outside of New Zealand?

We showed our footage to Chris Helgin. I was expecting him to pour scorn on the terrestrial-hunting scenario, but he became extremely animated and was genuinely excited by what he saw. His instinct was to give credence to Johnny's hypothesis, but with the proviso that he needed more information before he'd get too carried away. Determined to document the phenomenon in more scientific style, we went out to the site every remaining night we were in the crater hoping to come across our bats again, but sadly they never again showed their grisly faces. They would just have to remain one of the many mysteries of this extraordinary place.

After the excitements of the first few days, the sunlit wildlife encounters petered out as creatures that had once been brazen and inquisitive started to shuffle further away from these strange invaders into their world. After that, it was at night that most of the really mysterious creatures showed their faces. They were certainly a magical hyper-real experience that none of us would forget.

Several hours after nightfall and a bully beef stew we had to battle to keep down, we headed out in a mean drizzle, first several hundred metres upstream through the forest, then a steep scramble and slide down to the main river. As we reached it, all of a sudden a tiny Cruise Missile came out of nowhere, aiming straight at my head. I ducked to the side like a boxer dodging a jab, and it swooped past my ear with an audible rush of air through feathers. Within seconds the whole team was being dive-bombed by fluttering white shapes. They were Gordon's flycatchers, all three of them buzzing within feet of our ears.

It seemed that they were nervous and overprotective because of the presence of their youngsters, and spooked by the first torchlight they'd ever seen. This mobbing behaviour is one of the most remarkable feats of altruism you'll see in the animal kingdom, and a dazzling display of bravery. Each bird would take it in turns to run sorties across the river, hurtling towards our heads, flying straight at our faces before banking away at the last millisecond. Though they are scintillating avian athletes, there is no doubt they are putting themselves in extreme danger in driving away a threat many thousands of times their own size. Of course, you cannot really ascribe human emotions or nobility to the intentions of animals: what they are doing here is bombarding a larger, slower animal in order to ensure the furtherance of their genetic material, contained in the neat package that is their chick. Similarly, there is no provable valour in the vigorous mobbing behaviour exhibited by certain invertebrates, such as purple emperor butterflies found in the UK, which will happily chase away other butterflies and even birds from its territory.

They are all acting from instinct, spurred on by the simple programming of their primitive brains. Altruism is a strategy that has evolved because it works, because some animals that exhibit altruistic behaviours succeed over their competitors. However, it is very difficult *not* to imbue animals with human sentiments; certainly, it is hard to hear blackbirds' mournful 'bereavement' song, which they sing upon losing their chicks, without thinking that they are indeed sorrowful, lamenting the loss of their chicks as any human mother would. Our flycatchers, too, have remarkable personalities. One bird swooped high, and hovered almost like a hummingbird above me, then launched repeated attacks at my head before retreating to a perch on a rock in the centre of the stream. Another flew direct, unwavering, furious single runs across the river, zipped past and then headed for vines overhanging the water. Even the rain could not dissuade them from their course.

We waded across the river, shielding the camera from splashes from below and from rain and dive-bombing birds from above, and made our way up a narrow feeder stream, which flowed into the river at right-angles. As we followed the stream it quickly became a gorge narrow enough that you could spread your arms out and touch the rock walls on either side. The gorge made its way up a series of small waterfalls. Mossy boulders wobbled and tumbled under our feet. This streamway turned out to be a unique place for frogs.

The first was a tree frog of the *Nyctimystes* group, and distinctly different from all the others we'd been finding outside Bosavi and almost certainly a new species. He sat with his throat pouch pulsing at the streamside. He had gigantic eyes, and if he'd been scaled up to person size they would have been great brown footballs in his head. The way you identify this particular species from other tree frogs is from the pronounced palpebral venation, which is the scientific way of describing a miraculous feature of the eyes. One eyelid, which is covered in a fine latticework of veins, blinks up from below to cover the eyeball. The eye now looks like a stained-glass window, or fine filigree lacework. It is quite extraordinarily beautiful.

That first frog we came across that night was a real personality, and only further cemented my deep love for these charismatic beasts. After struggling briefly, he clearly decided that my hand was as good a place as any to sit and wait for a mate, and he settled himself till he was comfortable. Eventually, though, I'd said all I could about him to camera, and really wanted him to hop off to safety, so gave him a little nudge. Off he hopped behind my head. 'Where did he go?' I asked, looking up to see the crew wearing big silly grins on their faces. 'He's sat on your shoulder,' they replied. Not only that, but after landing he'd turned right round so he was facing the same way as me, looking for all the world like a little Kermit parrot sat happily on a pirate's shoulder. I took him back on my hand, and aimed him towards a suitable-looking branch. But he had other ideas, and hopped straight into my face and proceeded to climb over my nose and ears.

I'd been set the task of collecting some of the Bosavi frogs for the science department, but would sooner have cut off one of my fingers than remove this wonder from his home. I did, however, manage to collect eight different species in this one stream in just half an hour.

I had little doubt that most of them would turn out to be new to science. One major failure, though, for which I will still be kicking myself on my deathbed, happened up the streambed after many hours of soggy searching. My torch beam caught the ruby glow of a large pair of eyes, and behind it the lurking shape of a massive frog. To give you an idea of how massive, it had to be held in two hands, the legs draped down nearly a foot long, the body was the size of an old-fashioned milk bottle, and it had a mouth that could have swallowed all the other frogs whole. Steve Greenwood was beside himself with excitement, but its coloration was a dead ringer for the New Guinea species I'd come to think of as the common-as-muck frog – and I simply dismissed it as an item of no importance and returned it to the streambed, though with the nagging feeling that I was doing something wrong.

Later, when we met up with Allen 'Kermit' Allison back in Moresby, he described to me the creature he would most have wanted to find in the crater: 'Well, uh, the locals talk about finding these huge frogs in the streams, with a sort of quack or cluck call, they call them jungle chickens ... now that's exciting, sounds like the biggest frog in New Guinea, and I guess it must be a new species.' I bit my lip, and didn't say a word.

17

Dark-brown water thunders into my face, taking my feet from beneath me and the breath from my lungs. I almost laugh to myself. 'How stupid! How could I lose my footing in this little river?' But then I surface, and find myself looking downstream. Just below me the river roars into a narrow gulley with vertical walls. I'd swum down this gorge several times. What I see makes me cling to the safety line, horrified. Inside the gulley the water has turned into a crashing black vision of Hell, a giant washing-machine tumbling small trees and turning them into matchsticks. There is no doubt that to lose my grip on the line now would mean a messy, horrid death. How did it come to this? Just hours before, the river was our succour and our saviour, but the raging rains have turned it into a monster. Now it pulls at my legs with phenomenal force, a thousand roaring demons trying to drag me off to Valhallal. This would be a very bad way to go . . .

The constant rain over the last few days was transforming Bosavi back into the horror show we were expecting when we first made our way in. There was no escaping the wet. The only place that offered any comfort was under the locals' tarpaulin. They had carpeted the floor with bracken as soon as they arrived, and were walking inside with bare feet. Inside was dry and cosy, though the constantly burning fire made it unbearably smoky, rather like an East End pub before the smoking ban came into force. The fern carpet seemed such a fine idea that I headed off with my machete and collected enough to make myself one. The sweet smell of fresh-cut ferns transported me back to my childhood, making camps with bracken roofs and floors and cutting my way through the wide expanses of ferns that filled the woods in summer around my home in Surrey.

Bracken is the most successful of all the ferns: fossil records show the plant is at least fifty-five million years old, and this same species here on the hillsides of New Guinea is found on every continent bar Antarctica. It's the very one I'd known as a child. As I chopped great wads of it down in this extinct Melanesian volcano, the tart green smell of chlorophyll filled my nostrils.

I'd love to report that my own fern carpet was as great a triumph as the one in the locals' tent. Certainly, for the first hour or so it was a joy to swing my feet out from inside the hammock and drop down barefoot on to the soft fronds. It gave me a place to treat my suffering feet – as well as somewhere to sort and organise my stinking kit. For a few brief moments I thought I might have stumbled on something special that might transform my jungle experience, but soon Steve, Nick and Jonny wandered past in their muck-caked wellies and stopped for a chat, stomping black mud into my nice clean carpet. I was more upset than if someone had trodden dog poo into my rugs back home.

With the huge amount of water thudding into the crater, the river was undergoing a complete metamorphosis, its emerald waters raging brown and getting so powerful that it was in most places uncrossable, and in some extremely dangerous.

The incessant downpour meant that everything in camp was plastered with mud. Protocol of an evening was to finish your night walk, eat and sort out camera gear while still wearing your soaking-wet clothes, and then when everything was done to retire to your hammock, and only at the last possible minute change into your dry clothes and slip into your sleeping bag. We were at a thousand metres above sea level in the crater, so this routine was even more important. Night-times were fresh, even quite cold if it got windy. To have a dry set of clothes and sleeping bag was vital if there was to be any chance of getting any comfort or sleep.

One evening, when we'd managed to get the generator to work and all the hard graft was done, Luke decided to play a movie on his laptop. Not surprisingly, within minutes we'd all been crowded out of view of the tiny screen by the local guys as they pushed in closer. The chosen film was the first in the *Lord of the Rings* trilogy, and though they didn't understand what was being said they were nonetheless swept along with the story and thoroughly enjoying themselves. I've never seen quite such a dramatic reaction to a movie, with belly laughs at the funny bits, shrieks at the scary bits, and cheers at the heroism. However, as the film progressed and the hobbits pushed into the mines to be attacked by orcs and trolls, Maxi grabbed his brother Chocol in terror, squealing and then hiding his eyes from the screen. It was as if a three-year-old was watching *The Exorcist* and hugging his mother to avoid seeing the gory bits! The older men became very serious and muttered amongst themselves, pointing at the screen, while the younger ones yelped and whimpered. At the end, they all sat around solemnly discussing what they had seen.

Finally Maxi piped up, as if seeking the answer to a great secret. 'Where is this place, Mr Istiv?' he asked.

'It was filmed in New Zealand,' I said. 'Go very far south and east from here and you'll find it.'

'And those short men, with the beards, they are only in New Zealand?'

'The dwarfs?!! No no, Maxi, they aren't real, they're just made up for the film – it's not real!' Chocol translated, and they all nodded very seriously and passed this new piece of information around.

'And these ugly men, with the bad faces, they are very frightening men. Do you find them in your country?'

'What, the orcs?! No, Chocol, honestly, it's just a film, they're not real!'

He nodded his understanding, and they began talking amongst themselves again.

'It is very interesting to us, this place, but I think it is better here on our mountain.'

It astounded us that they hadn't grasped that this was a fantasy world – and yet, *was* it so surprising? When we first arrived here, they would stare at our books and photos with wonder, grabbing them from our hands so they could gaze at the pictures and touch the faces, reverentially. However, soon they learnt the (to them) inconceivable truth that still photos were just the beginning, that we had big cameras that could capture reality and put it on to a screen. After having been dragged into the twenty-first-century world of Western miracles and bizarre white man's cargo, the next thing they had discovered was that out there in this big world they knew nothing of was a place where dragons and wizards battled each other with swords and arrows. Thinking about it, though, this was probably less of a cultural wake-up call to them than if we'd shown them films about the real modern world.

Luke relayed a story about how one of his colleagues, with a kind of irresponsibility I find incomprehensible, had shown a pornographic film to a group of Papuans, evidently because he thought it would be funny to see how they reacted. The men in most Papuan cultures treat sex merely as an irregular and almost ceremonial process, and hadn't known what to do with this new information or with the arousal of desire in themselves. Luke reported that not only did they nearly come to blows with the local men when they tried to steal the laptops with the films on them, but the relationships between the men and women of the village were thrown into turmoil. The men beat any women who didn't acquiesce to the new sexual practices, and the women beat the men who had, presumably, tried alien lewdnesses with them that sat uncomfortably with ancient Melanesian sexual traditions. It sounded as if that one movie showing nearly brought the village to its knees, so to speak.

While I would never be so crass or stupid as to do something like that, our time here in the crater with the peoples of Sienna Falls and Iggycelabo did give me pause for reflection on the impact the presence of people like me makes around the world. On a recent expedition to the Himalayan kingdom of Bhutan I passed what seemed like a significant milestone, as it marked the one-hundredth country I've visited. It's a bit meaningless, really: in order for a country to qualify, you only need to have slept a night there, so I'm including Abu Dhabi and Dubai where I merely stayed in a hotel when flights were delayed. I'm counting all the countries of the UK separately, and I'm certainly including Tibet, no matter what the occupying Chinese government says.

Perhaps it's egocentric to think about it like this, but it was a milestone nonetheless. It's also a natural time to look back on my travels, doubtless embarked on for selfish reasons, and ask what I've achieved, or more relevant, what damage I may have done.

When I first started backpacking, I took pride in travelling well. I stayed a long time in each country, made a massive effort to learn the language, and did everything possible to respect the people's customs. I treated other Westerners like lepers, and lived amongst remote peoples as one of them, ensuring every penny I spent went straight into the pockets of the people who really needed it.

Admittedly, in retrospect I must also have been a repulsively smug travel snob.

I'll never forget as a nineteen-year-old coming back from four months in the most far-flung reaches of eastern Indonesia, much of it spent living with an animist tribe who had never before seen white people. I'd drunk still-warm buffalo blood at their funeral rites, prayed over the recently dead corpse of my host's daughter, lived on nothing but rice and rock salt, and pretty much fancied myself to be the greatest traveller who'd ever lived. When I got back to Bali, I took myself to a travellers' bar and accosted the first foreigner I could lay my hands on – an attractive British girl – and proceeded to unload what I saw as the greatest adventure story ever told. Some while later, she yawned, then headed off to bed as she had 'an early flight the following day'. Just an hour later, bumbling around looking for someone else to entertain with my tales, I saw her in another bar laughing with a group of backpackers. Ouch. Lesson learnt. You can take yourself way too seriously when it comes to this travelling lark.

But nowadays I sometimes fear I've gone too far the other way. As a television crew, we may strand ourselves in the jungle sleeping rough for months on end, but when we get back to civilisation it's straight to a plush hotel, probably a multinational chain that gives little or nothing back to the local community. Furthermore, the hotel may well have been hacked out of virgin mangroves or stand on a plot of village land that was bulldozed over to make room for it. We eat meals that would cost more than the villagers we've been living with might earn in a year, swim in the pool, surf the Internet and act like tourists the teenage me would have disdained. There's also no rule that ensures that a camera- or soundman sent out to film 'first contact' trips with remote tribes possesses any kind of sensitivity. I've worked with plenty of big metropolitan production crews who spend most of their lives filming celebrity make-over shows or MTV music videos, and wouldn't have a clue how to deal sensitively with remote cultures and environments. But that doesn't mean they won't be sent out to make films on them anyway.

After the tremendous success of the documentary *Tribe*, showing the respectful and self-effacing anthropologist Bruce Parry living side by side with peoples whose lives are very different from ours, British television now abounds with programmes in which proud people and their traditions are seen as a freak-show filming opportunity. There are shows in which arrogant, moronic teenagers are sent to ancient tribes to attend rituals with a view to losing weight, take their mind-altering substances as if at some jungle Glastonbury, or to discover themselves, or just to have a bit of a laugh at the guys in the penis gourds.

There have also been series in which clueless young sportsmen take part in tribal games and challenges as if they were circus attractions – regardless of the fact that such events are of great cultural significance – then proceed to kick the living daylights out of the smaller local peoples while the villagers look on in bemusement.

There's no law saying television *has* to be sensitive to its impact on local peoples, just as there's no law saying those *Jackass* morons can't dry-hump an endangered species or leap on the back of a great white shark. Also, though I like to think my trips are undertaken with understanding and intelligence, none of us are immune. On a recent trip to Madagascar, my cameraman spent every second he wasn't filming watching movies on his laptop, ignoring the wondrous world rushing by outside the windows of our air-conditioned minibus. These are times when I feel thoroughly ashamed of what we have become.

And what of the effect on the peoples who have taken me in as a son or a brother over the years? People like Maxi and Chocol are living in a tradition that is thousands of years old. They own little or nothing, but are genuinely happier than most people I know back in the Western world. However, after he'd had a taste of our riches, Chocol, particularly, experienced desire, and desperately coveted our generators, laptops and satellite phones. And who am I to suggest he shouldn't have them? Wouldn't it be horrifically patronising to say: 'Hey guys, trust me, you'll be far happier living a simple life. Don't worry about getting rich like me, it won't make you happy'?

That doesn't stop me from wondering if my presence here has done more harm to them than good. But what do you do? Should you just stay home, or only go to places that've already been spoiled by Western tourism? Well, I don't think so, and if only because thoughtful travelling could offset some of the wrongs perpetrated by the imbecile travellers and film-makers. Every traveller, whether they're on a weekend break or a year's wild wanderings, should have an ethos, and

I'd like to suggest the Hippocratic maxim 'First do no harm.' Ultimately, we travel for selfish reasons, but we should be educated enough to think carefully about the legacy we might leave behind. Travel is so easy nowadays that anyone can get to places that are as yet unspoilt, without the knowledge and experience it would usually required to get there. As long as I'm not the Australian surfer lashing out at an Asian fisherman who tried to hold his hand (I've seen it happen), as long as I'm not wearing a bikini or having sex on a beach in an Islamic country, as long as I'm not the one showing pornography to Papuan tribesmen, as long as I'm not the one actually doing the spoiling ... perhaps I can make do with that.

The next day, when we were lazing around camp letting our meal of rice and boiled bully beef digest, Tom, one of the 'uncles' from Sienna Falls, came into camp bearing a wriggling pillowcase. Until then Tom had been most remarkable for the fact that on day one I'd caught him putting *six* heaped teaspoons of powdered milk, six of coffee *and* six of our precious sugar all into one glutinous mugful, barely leaving room for the hot water. However, he was now about to become something of a hero. Out of his writhing pillowcase emerged one of the cutest-looking creatures you could ever wish to see. It looked very much like a slow loris, with huge brown eyes that fixed you without any suggestion of sentience but were almost alien in the impression they gave of staring right into your soul. He walked carefully up my arm, his claws scratching me red-raw, the thick pelt that kept him warm in this wet montane climate silky beneath my fingers. I could tell from the look on Chris's face that Tom had found something special.

'OK, so this is a cuscus, it's a marsupial, which means it rears its young in a pouch,' Chris said. And with that he opened a small vent in the fur between the rear legs and looked inside at the pink skin: 'Though this one has no young at the moment.'

These cuscus are strange, almost eerie creatures, about the size of an overfed house cat, with a prehensile tail to help with climbing and a thick woolly coat that in some species comes in burnished gold with ivory and ochre spots. Apart from when mating or raising their young, they are known to spend their lives alone, and can be quite fierce with rivals and interlopers.

Next, Chris opened its mouth and looked inside at the teeth. You could have cut the tension even with our bluntest machete.

'Yup, this seems a good match for the skull Jonny found down in the river the other day.'

But before he could commit, Chris took more measurements, weighing and assessing the dimensions of the cuddly critter, and compared it carefully with the photos and diagrams in his books and on his computer. All the while, the calm cuscus wandered over me as if I was a climbing frame, perhaps a bit dopey as this normally nocturnal creature was awake during the day. Its soft pelt smelled strong, but not unpleasant, against my face.

'I can't even begin to describe how it feels to have an animal in my hands that is this beautiful, and in all probability has never before been seen by science,' I said.

And then Chris uttered the magic words: 'I think what we have is a cuscus that long ago has been isolated on this volcano, has not been able to have any contact with any of its relatives, and has become something here, in its isolation, that is unique to Bosavi.'

The reactions from the rest of us were noteworthy. It was as if the blanket of rain clouds over the crater just fizzled away. We had our result; a new kind of mammal!

'He's just totally chilled out, just sitting here in my arms, he has no idea how important he is! Little guy, you're a major scientific discovery.'

I travel the world looking for new species in many different places, and we find them, we do find new mammals, but they're usually rodents or bats. To find something, a marsupial, an animal that's this size, is cause for major celebration.

'Yeah, crack open the champagne – or crack open the bully beef!'

While you can smoke out any tropical rainforest tree and find new species of invertebrates, and while Allen turned out around sixteen species of reptiles and frogs on the trip, to come up with a new mammal like this was phenomenally rare, and would be a genuine scientific coup.

Not only that, but once we'd found one, they started turning up everywhere. I'd step out from behind the dripping tarpaulin we called home to answer a call of nature, and catch the red eyes of the Bosavi cuscus in my torchlight just feet away. At night I was even woken up by the heavy, musty smell of one right overhead, and after managing to film three in one tree, including a pair grooming each other, we began to doubt the received wisdom about them being solitary animals. Eventually they were turning up so frequently we just stopped filming them. 'What's that?' – 'Oh, nothing, just another of our new species.'

When Alfred Russel Wallace arrived in New Guinea at the end of the nineteenth century, he noted: 'The Mammalia yet discovered are only seventeen in number.' Today the checklist numbers just less than two hundred and fifty species, and there are surely more to be discovered. In his masterpiece *The Malay Archipelago* Wallace comments: 'The unknown portion of this great island is the greatest *terra incognita* that still remains for the naturalist to explore, and the only region where altogether new and unimagined forms of life may perhaps be found.' Here in Bosavi, having been party to the excitement following the discovery of our new mammal, it is hard to imagine how early naturalists must have felt here. They must have woken every day from dreams of discovering strange mythological beasts, dinosaurs and other animals that would take the whole world by storm. What a time that must have been.

In some ways, however, biological science has not moved on at all from those days, and in ways that my colleagues and I found rather upsetting. The unpalatable truth is that even in today's high-tech world, where the full genome of an animal can be mapped from simple blood samples and huge databases on species have been set up, the process of declaring a new species is still Victorian. In order to prove what you have is new, you need a holotype – that is, the first specimen of the species to be described. Ideally, you would have several, perhaps many, of those first specimens, known as paratypes, to put together a type series that more completely shows differences across the ages, sexes and individual physical differences. Each one of these specimens will have been killed with ether or by a whack to the head, then the pelt and skeleton retained, to be kept in a museum drawer somewhere as part of their collection.

Now, I have to admit to a little naivety and sensitivity on this issue. I had no idea the process still went on in modern biology until our

Guyana expedition. We had a bat scientist along, who we christened Doctor Death. He caught tens of specimens in his mist nets every night, which he would spend the next day photographing, then killing and stuffing. And he didn't just do it with rare and unusual species, but with the common ones too, if he didn't happen to have them in his museum's collection. Some of the film crew would go out just before Dr Death at night, and untangle bats from his nets to help them escape to freedom. We all wanted the glory of a new mammal discovery on the expedition, but nobody wanted animals to die for the sake of it.

Here in Bosavi this feeling was even more powerful, and probably because the animal in question looked so incredibly cute. Now, I understand this makes me a sentimentalist, but I don't mind. Surely in today's world we could just take an epithelial swab, a DNA sample and a sequence of photographs? Why do we have to kill the animals? I totally understand that it's necessary with invertebrates, whose sexual organs may have to be dissected under a microscope for the new species to be described. They are, additionally, much more numerous, and have brains half the size of a full stop. But why should we have to do it with mammals? And more importantly, what are the implications for the populations of these animals if we lay waste to them in the name of science?

All my experiences in New Guinea had shown me that mammals are nowhere here especially numerous. Chris absolutely refuted the suggestion that we could be making any impact on the populations of our cuscus by hunting them, but there are many species around the world that are so thin on the ground that collecting five or six individuals could push them to extinction. Famously, the last two remaining great auks – a flightless bird that was not unlike an Arctic Circle penguin – were killed on an Icelandic island for a zoological collection, and Steller's sea cow (a nine-metre-long ten-tonne manatee) was hunted to extinction within a few years of its discovery by the eighteenth-century German naturalist Georg Steller.

After a few days of collecting and preserving our adorable cuscus, Jonny, Steve and Chris had a tempestuous argument, as Steve and Jonny tried to convince Chris that he had enough specimens and we should just let further potential subjects go. Jonny and Steve are both proper scientists in their own right as well as film-makers – Steve an entomologist, Jonny a PhD zoologist – but even they were uncomfortable with the ways things were going. Chris countered that we had got him here under the pretence of doing real science and were stopping him doing his work. It was an uneasy working relationship for the remainder of the trip.

The following night Steve, Luke and I headed back to the creek in heavy rain, our headlamps turning each dancing raindrop into a plummeting firefly. Jurassic ferns dripped generously down the backs of our necks, slight slopes became amusing slides, steep slopes became possible death traps.

No sooner had we reached the riverbank than our flycatcher friends returned with a vengeance. I had kind of expected that they'd get used to our night-time forays and decide to save their energy, but in fact their attacks increased in ferocity. The male, particularly, dive-bombed us like a Japanese Zero around an American bomber. His tiny head crashed into my stomach, flew between my legs and stopped to perch no more than a foot from my face, staring at me with one angry black eye before hovering away like a white hummingbird, chirruping and calling in the youngster and the mother as reinforcements. They had all the bravery of wrens mobbing a buzzard back home, and it had the desired effect. I ended up flat on my backside, spinning around in an effort to follow him with the torchlight over the slippery rocks, and Luke got so fed up trying to film the little whirling dervishes he very near had a major tantrum. The male bird not only tracked us over the river, but even followed us up the creek, clearly calling back to his family: 'Look at this, guys – they're running away!'

Just up alongside the stream, my attention was caught by patches of luminous fungus that adorned the collapsed rotten tree trunks. This bioluminescence serves to attract small flies and moths, which land on the fungi, pick up their spores and then unwittingly disperse them – the wonderful tastes and smells of truffles serve the same purpose. This natural light was truly stunning, though we could only see it when we turned off the lights and let our eyes get adjusted. Even then the effect was only faint, but it appeared as if all the dead trees had been painted a ghostly, glowing green.

These decay fungi are incredibly important, as they help massive trees to rot, thus returning their nutrients to the soil. Another of the wonders of fungi was discovered on the base of a tree nearby, which sported large brackets called 'conks'. These look rather like fat brown frisbees that have been thrown at the trunk and then embedded themselves in it. They're so strong and tough that you could imagine using them as a ladder to scale their host tree. These particular brackets were releasing their spores, and millions of them were drifting lazily skywards like smoke from a smouldering bonfire. It was strangely beautiful when lit by our white filming lights, the spores drifting up and off to make new fungal life elsewhere. Some bracket fungi can live to be seventy years old, and liberate thirty billion spores a day for six months, that's five trillion spores in one year!

Again, the creek was a panoply of frogs, even more wondrous because there were different species in evidence from the night before. One micro-treefrog continued calling even when we got inches away from him, his throat pouch and lungs expanding like overblown balloons. The quacks he and his kind made turned both the river and the camp into a weird duck singalong, while other species produced repetitive, ratcheting clicks, or chirps, burps, or buzzes like a vibrating mobile phone on a table.

One of my main missions was to try and find a snake inside the crater, as none had been spotted since we got here. My ideal would have been a Boelen's python, a stockily built snake that may reach three metres in length, coloured iridescent gunmetal-purple with white trim. They occur above a thousand metres but below the tree line, but are fabulously rare and little known. The 'uncles' all agreed that they'd seen snakes here matching that description, but sadly the cool weather and the rain would certainly keep pythons curled up in their retreats and burrows, so my search was doomed to failure. In fact, I never saw a single snake in my entire time in Bosavi, though Nick swears he saw one of the small-eyed variety while up in the forests at night, slithering right past his feet while he was recording frog calls.

As I've intimated before, the big problem with expeditions in the rainforest is that the dense foliage forms such a thick cover around you that you rarely, if ever, have a view. Essentially, in the day we were entombed in green, and at night in a dark chamber that was only minimally – and tantalisingly – teased apart by the beams of our headtorches. This particular night, though, we pressed on and on and ever upwards, to the highest point we'd reached since our entry into the crater, following a ridgeline that offered us an unusually achievable mossy path. Two hundred and fifty metres above our camp, one green wall of the ridgeline abruptly disappeared, and we were afforded a rare and spectacular view.

As the clouds briefly parted and the moonlight broke through, we looked towards the southern wall of the volcano and were reminded – as if we *needed* reminding – that we were truly in a forgotten place. The volcano towered above us, a near-vertical natural fortress a thousand metres high, its rim jagged against the moon-washed sky. Far below us, the shiny white saucer of the reflected moon danced on our river, catching the billion sparkles of the waterfalls. For a brief moment I was a sword-bearing hero in a fantasy world, a fur-clad Cro-Magnon on the trail of a sabre-toothed cat, a Victorian naturalist searching for a forgotten dinosaur.

Had anyone ever looked out on such a wonderland before? It was so beautiful, so deeply timeless, that primordial tears sprung to my eyes. But then the clouds rolled back across the moon and the moment was lost, leaving me feeling faintly silly to have been so affected.

We turned tail for camp, stopping to scan the deadfall logs and saplings for orchids, usually to be found in the characteristic bunches of leaves and stems. There are believed to be near thirty thousand species of orchid around the world, but cool, damp tropical forests like here in Bosavi are where they are at their most numerous. Most were growing as epiphytes with aerial roots, some of them parasitising nutrients from the plants they were attached to. It seems all members of a species come into flower at the same time. Only one species was in bloom that night, a thimble-sized bright-red flower, several to each plant. They were stunning, and seemed to be everywhere.

Beneath the roots of the trees on our ridgeline the ground had been excavated into networks of tunnels. Around the openings, masses of waiting leeches told us the culprit was a mammal – namely, a long-beaked echidna. This is one of the most unusual beasts in a world packed with weirdos. They're kind of like a cross between a big black mole and a hedgehog, with an anteater's snout and long digging claws, and they waddle along the ground on short, powerful legs. With the platypus they are the only remaining monotremes, which means 'single hole' and refers to the cloaca, their single opening for excretion and sexual activity. The monotremes are like other mammals in that they feed their young on milk, have hair on their bodies and a high metabolic rate, but unlike other mammals in that they lay eggs.

Once the eggs have hatched, the young (in the echidna's case, known as 'puggles') crawl into a pouch, where they lap the milk, secreted through the mother's skin rather than via a nipple. It's generally considered that the monotremes represent the last remnants of an especially ancient group of protomammals, which split from the other marsupials and placental mammals perhaps a hundred and fifty million years ago. If disturbed, they'll roll up into a ball, with their long spines preventing all but the most determined of predators getting to their soft underbelly. Unfortunately, this defence doesn't deter humans, who simply pick the echidnas up and carry them off to be eaten. Human hunting and habitat destruction are the sole reason that echidnas are now endangered throughout New Guinea, but the short-beaked echidna in Australia is better off as far as its conservation is concerned.

None of the team had ever seen a long-beaked in its natural environment; the only time I'd encountered one on my New Guinea trips was in the market in the Baliem Valley, being sold for meat. I had, however, come across several short-beaked in Australia, and found them to be quite the most charismatic, charming beasts. We were determined to find one for our film, though tacitly had all agreed that if we did, we'd keep the discovery from Chris or claim that the echidna had escaped at the last moment. Gordon's team and my team went our separate ways on the echidna hunt, the first lot possessing the distinct advantage of a thermal-imaging camera, which picks up the heat that's generated by moving animals. The image produced is rather like that in Hollywood's *Predator*, with cool trees and background shown in blue, and moving mammals red and orange at their peripheries and almost white-hot at their centres. In addition, they had Uncle Tom, who was turning out to be the king of mammal hunters, and rarely came back from a night walk without something warm-blooded in tow.

Several days into the competition, and both teams had headed up on to the same hillside. After an hour of fruitless searching, we saw the lights of Gordon's camera a short distance up the hill, and headed up to join them to see if they'd found anything. Gordon had the thermal camera, and was scanning the undergrowth.

'Hey Gordie, how are you getting on with the cheating camera?' I asked.

'It's great, but very difficult to get any sense of scale. I thought I was filming Jonny just a second ago, then suddenly he stood up and jumped about the height of a house – it turned out it was actually a mouse.'

'This really isn't a fair fight, is it? My team has to just rely on our wits and superhuman tracking abilities to find anything at all. You boys just let the technology do it for you. It's laziness, if anything, but I guess if you need it . . .'

I'm obviously not even remotely bitter or competitive about all this.

'We got some truly remarkable footage on it earlier, Steve – you'll love this'. Gordon wound back through the stuff he'd recorded to show me his discovery. 'Ah, here we go – what d'you think that is?'

He shows me a blue background, the trees and plants subtly different colours from the ambient blue. Slap bang in the centre of the shot was an incongruous orange and yellow mound, about mammalian body temperature though like no living creature I've ever seen before.

'What the hell is that, Gordie?'

'I'm not sure, but I think it might be a new faeces.' Gordon and Jonny are trying not to crease up with laughter.

Gordon confesses: 'I got caught a bit short and had to use the jungle bathroom – that's the results – you've got to love technology!'

'I'm sure the BBC will be delighted you've been using their high-tech kit for such serious ends, Gordie.'

'Well, we've got a few other things on there too. There's a forest mouse and – well, actually there's just the forest mouse.'

At that moment Uncle Tom walked quietly into our group, bearing a gorgeous furry bundle just as you'd carry a newborn baby.

'Wow, look at that!' Jonny exclaimed. 'It's one of those coppery ringtails!'

He reached down to see if there was any sign of young, and there snuggled inside the pouch was a tiny, naked blind baby, attached to the teat.

The coppery ringtail is a gorgeous little possum, similar in appearance to the cuscus, but smaller, with large dark eyes and a longer tail. It looked as defenceless as a kitten. As I'd not had any successes on camera for several days, I decided to take the possum off Uncle Tom and do a piece to the camera about it. Tom had it cradled face up, like a baby, but clasped firmly behind the neck and hind legs, so I tried to take it off him as you would take a proffered baby. This was a big mistake. As soon as I took it from Tom's hands, the adorable kitten turned into a spitting, snarling wildcat. Its furious sharp claws ripped through the palms of my hands, cutting slices as deep as a carving knife. I yelped with pain and surprise, and instinctively clasped her to me in order not to drop her. To reward me for this protective gesture, she bit straight through my T-shirt and into my chest until the teeth of top and bottom jaw met. It was like having a pair of white-hot knitting needles piercing my pectoral, and it was all I could do to stop from screaming like a baby.

'Wait, Steve, I haven't got the camera up to speed yet.' Robin was struggling to get the filming light ready. 'Just hold it there, mate.'

'Get it off me, JK, get it *off*!'

Johnny was biting his lip trying not to laugh, but tried to take the ringtail by the scruff, while Uncle Tom attempted to disengage the firmly latched teeth. This was now a matter of real embarrassment. All I could think of to get the damn thing to stop consuming me like a chunk of chorizo was to offer it a nearby tree. As soon as the sapling was within reach, the little beast grabbed the bark, then released its teeth and scrabbled at an alarming pace into the canopy.

'*No*, Steve! I didn't get any of that – get it back!' Robin is ready to roll, but our ringtail is now no more than a bloody memory.

'Little bugger's sliced me right up. I can't *believe* the worst wound I've had out here's from a cuddly little possum! Please God, don't tell anyone about this!'

'No way – God, Steve – of course we won't!' Gordie replied, with a sincerity that told me he'd be on the sat phone to BBC World News about it before the night was out.

'I guess that's one furry mammal Chris won't get his hands on,' Jonny quipped drily.

'Yup, my ploy worked an absolute treat.'

Rather embarrassingly, this little episode meant my team had to retire to camp, as any animal-inflicted wound had to be well cleaned and treated. Jane was over the moon to have something to deal with, though had to battle to contain a smirk when she heard I'd been savaged by a cuddly toy. *I* was certainly not grinning, though, when she scoured the wound out with iodine and a scrubbing brush. Protocol warns against stitching animal bites, as this can internalise unpleasant bacteria; otherwise, I'd probably have had a couple of stitches. And I do most of my work with venomous snakes and crocodiles. Oh, the shame of it.

Returning at around eleven, we'd no sooner put our cameras into the hotbox than Gordon strode into camp, his crew in tow, filming him as he approached us. He was soaking wet and bedraggled, but with the biggest smile on his face imaginable. He clearly had no idea how to tell us something of the magnitude he'd been storing up. He'd probably planned to say something funny or nonchalant, but in the event it just burst out.

'New species of giant woolly rat!'

I just stood there with my mouth open. 'No! That's unbelievable – are you serious or are you making it up?'

Gordon held his hands apart to show the creature's length, like the proverbial fisherman describing the one that got away. His hands were a metre apart. Then he halved the length, more or less. 'I guess the body must be this big, about the size of a fat house cat.' Then his hands go back to the original length. 'But the tail is probably even longer than the body, so in total – well, it must be this big!'

'Was it in a trap?'

'No, it was just running around the forest, tame as anything!'

'No way! Have you shown Chris?'

'Yeah, he got there just after we found it, he reckons it's the biggest rat in the world.'

'No way, no way! And is it?'

'Yup, Chris is sure, it's a new species – the Bosavi giant woolly rat.'

'Aaaaarrrgggghhhhhh! Gordie, I don't believe it, that's unbelievable!'

Again, Gordon just couldn't hold back. He was ecstatic. The earlier competitiveness between us was forgotten. This was a triumph for the whole team, for the whole expedition, the whole series.

'You've got to see it, Steve. It's genuinely the largest rat in the world, it's the size of a fricking dog!'

There were hugs, high-fives, handshakes and back-slapping all round.

'So where is it? Chris hasn't offed it already, has he?' I asked. We all desperately wanted to see the beast that had brought us all here.

'No, no, I just – you won't believe it, they've got it in a rice sack out back – the thing's practically tame – it was just sat there in front of me, chomping on roots and things, didn't seem fazed by us at all.'

'So how the hell did you find it, then?' Steve Greenwood was beside himself with glee, and I wasn't surprised, because in one quick moment his whole series had turned into a roaring success.

'Well, we were up on that high plateau behind camp, we'd had a pretty good night, we'd already got one beautiful painted ringtail and seen our new species of cuscus a couple of times, but then suddenly someone shouted 'Echidna!' We went charging over and saw Uncle Tom grappling with something on the ground, but when we got close, we could see it *wasn't* an echidna.'

'It had to be Uncle Tom, didn't it?' I said almost ruefully.

'Yeah, he's *so* good. It's almost like he senses the animals – it's freaky.'

'So you saw it wasn't an echidna?'

'Yeah, and I was just about to give up, it's been such a miserable night.' It had – the rain hadn't stopped tipping down since lunchtime.

'And we were figuring it was "just another" cuscus. But then we got closer – and did a massive double take. I was sort of like "What the hell is that?". I mean, Uncle Tom had it by the tail, but it was totally relaxed, you know, grooming itself, eating bits of root, wiping the raindrops from its face. It was practically grinning.'

'Just like someone's pet rat.'

'Exactly! Like someone's pet rat that'd been on steroids, or something. It's just the biggest rat I've ever seen. Well, actually, it's the biggest rat anybody's ever seen, ever. It's the biggest rat in the world!'

At this, Gordon laughed at the ludicrousness of it all. This was to be the greatest achievement of our entire expedition, the one find that would guarantee us the headlines, shower us all in glory and underline our true purpose. Understandably, my Scottish colleague was beside himself with triumph. His team sat well into the night knocking back the last of the Scotch and retelling the tale over and over again. I'd be lying if I didn't admit to a little jealousy. Our team would have killed to be the discoverers of the Bosavi giant woolly rat, but now was a time for being gracious in defeat, and praising our colleagues for their success. It hurt, but the smoky Scotch helped to soothe the pain.

The next morning, I came face to face with our rodent friend as Chris took him out of the bag to measure him again and confirm his prognosis. This was indeed a very, very big rat. It weighed in at one and a half kilos, and from tail tip to the end of its pink snout was near a metre long. The thick white whiskers were probably as long as a 30 cm school ruler, and the long scaly tail was as thick as my thumb. If you took a photo of him and removed anything that gave a sense of scale, he'd look like something you might find going through your bins or scuttling off down a London Tube tunnel, but when you saw him with a person in the photo for scale, you simply wouldn't have believed your eyes.

Even after spending the night in the bag, he was quite calm in our company. Generally, a cornered wild rat is a ferocious adversary, a whirling dervish of teeth and claws. Having seen what a demure little possum could do when annoyed, I was in no doubt that this giant could have ripped me apart if he'd so chosen, but instead he just sat there nonchalantly. He had thick, deep fur – grizzled grey, which led us to assume he was quite an old specimen – to protect him from the mountain elements, and slightly milky eyes that told of cataracts.

The related woolly rats to be found elsewhere in the world are remarkably intelligent animals. In fact, some of our sturdy mascot's close cousins are trained to sniff out mines in war zones. Certainly, he seemed to have boundless personality. We all felt abominably disloyal when Chris – who had been standing in the background visibly sweating, and flinching every time the rat moved in case it made a break for freedom – put him back into his sack and took him off to his inevitable end. At least, we told ourselves, it was an old rat, near the end of its life, and the knowledge we'd take back from Bosavi could well be used to help save this place. We told ourselves that, but it didn't stop it hurting.

Finally, though, I felt that I had exorcised the curse of New Guinea. After so many years of fascination for the island that had listed towards hatred and fear, then back to fascination again, it seemed I had been a part of something worthy of the memory of Alfred Russel Wallace. That morning, the leeches, the disappointments, the frustrations, the pain from my fast-failing ankle all seemed worth it as we strolled around with grins on our faces. We were explorers, discoverers, finally worthy.

New Guinea, though, had one last hand to play before she'd let me leave.

As our departure drew near, the team plunged into gargantuan efforts to find more star wildlife. Our favoured location was across the river and up the mighty ridgeline that gave us our view and offered so much evidence of echidna activity. Late in the afternoon of that last day, in preparation for a big last nocturnal search, I took our remaining chunk of rope to rig up a handline across the flow, to steady ourselves as we crossed in the dark later on.

On one side was a huge boulder, and I wrapped the rope around it using a knot called a 'lorryman's hitch', which can serve as a sort of makeshift ratchet to tighten the line up. On the other side, I really struggled to find a suitable anchor spot: I had to find a notch in the rockface, then force a piece of sodden driftwood into the ground and loop the rope around it. It was far from ideal, but better than nothing at all. Finishing the safety line was the signal for the heaviest downpour we'd had since the first rains, and it continued way beyond nightfall. While we were consuming our nauseating bully beef, Lucky Luke and Steve Greenwood kept throwing glances across at each other: 'Surely Backshall isn't going to make us go out in this rain? We're not going to find anything in this!' But it was the last night, and there was no point saving energy or trying to protect the cameras any more. I was committed, and was not in the mood to listen to feeble excuses.

So the three of us went with Maxi across the river and headed up to 'echidna hill' in the pitch dark. We crossed the river without incident, barely even needing to use the handrail, then set off on the long uphill slog, drenched to the skin, the rain gleaming like falling fireflies in our headtorches. As we continued, the rain bounced off us and our kit as if we were passing through a giant carwash. Under normal circumstances we wouldn't have wanted to risk the kit and would have turned back, but we had several cameras still working and it was our last chance to find something miraculous. Driven by that familiar companion competitiveness, I pushed on. The rain dripped off the ends of our noses – we looked as if we'd been flushed down the pan fully clothed.

What I was mostly ignoring, though, is that most wildlife doesn't like the rain either. Reptiles, mammals and birds all hole up in heavy rains, becoming invisible even under the piercing beam of my spotlight. Even the frogs were near-impossible to find, as their songs were drowned out by the deluge. In over two hours of this wretched slog we found precisely nothing at all. Zero. It had been a spectacular anticlimax to the expedition, a freezing, draining waste of time. At our magic clearing there was precisely no view at all, and I decided to call it.

'OK, guys, I'm really sorry, this has been a total disaster. Let's head for home.'

The others breathed an audible sigh of relief. It was all over. All we needed to do now was retrace our steps, and in just over an hour we'd be back in camp, sipping Scotch before slipping into our sleeping bags. Not even bothering to scan the bushes and the ground for signs of life, our yomp back was fast and furious. Everyone had their heads down, alone in their own little worlds, beating the chill by moving as fast as possible, not stopping even for a second.

I, though, just didn't want to admit defeat, to admit it was all over. When we got to the creek where Tim had made camp for us on that first night in Bosavi, I left the other guys to head off back, and decided to spend a few hours in the creek to try and catch some frogs. After just fifteen minutes or so, with no results at all, the inevitable hit even me. I worked my way back to the river. Weirdly, the crew were still there at the riverside. Luke and Maxi appeared to be arguing, and only

Steve had crossed to the other side. My heart leapt – maybe they'd found something and had stopped to film it.

'What the hell's going on, Lucky? What you still doing here? Have you found something?'

'Mate, the river's up,' yelled Luke above the roar. 'I'm going to leave the camera on this side of the river under a tarp or something, it's just too dodgy to try and carry it across.'

'You what? Don't be daft, Lucky. I'll just drop my stuff on the other side then come back and give you a hand!'

This was bizarre. Luke is usually pretty tough. I guessed he was just a bit paranoid about being responsible for destroying the camera. The macho side of me wanted to show him quite how easy it was, and how lame he was being, so I strode out into the rushing brown water, holding on to my handline for stability. The water was a foot higher than normal and came up to my chest, racing along, but I got across and joined Steve on the other side.

'Luke's worried about the camera, they're just dithering on the other side,' I said to him, somewhat scathingly.

The response was not what I expected at all. Steve was looking grim. 'Oh God,' he replied, 'Maxi's terrified of the water – this could turn into a nightmare.'

'What? What do you mean, he's scared of the water?' I'd spent weeks with Maxi and never seen any evidence of that.

'Weren't you at basecamp when he nearly killed himself?'

I shook my head.

'Maxi was in one of the RIBs when the water levels were really high, They were carrying fuel drums, and one just shifted slightly when they hit a wave. Maxi was so scared that he just leapt out of the boat and into the river – and he couldn't swim, they had to rescue him. If he'd not had a life vest on he'd have drowned, for sure.'

This was all news to me, and not exactly good news, either. Maxi and Luke were stuck on the other side of the river in the dark. When I looked down I saw that the river was rising with every minute. It seemed to me that some affirmative action was needed, so I decided to go back across and lend a hand, perhaps to carry the camera or give Maxi a shoulder to balance on as he made his way across.

I dropped my rucksack, stepped back into the river – and was taken totally by surprise. In the few minutes I'd been talking to Steve the river had changed gear. From being just a bit of a challenge, the flow was now a full-bore raging torrent. No sooner was I knee deep than I was swept off my feet. I clung to the handline, at full stretch, and fought to pull myself back up to standing. I managed it, but then I was swept over again.

This couldn't be right – surely it had only been three or four minutes since I'd crossed over just fine, and now it was like trying to drag myself through a grade-four whitewater rapid. Every foot of distance across the maelstrom was hard-won. As I neared the other side I swept my headtorch downstream. Just below us the river narrowed into a gorge that I had swum through that morning. Now it was a crashing, roaring tumult of brown and white water, tossing fair-sized trees around as if they were nothing at all, then channelling into a black tunnel. What hell lay beyond I couldn't even guess at. I felt my stomach leap upwards. To be swept down into that dark place would have meant certain death.

I have experienced flash floods on several occasions in the rainforest, and every time they have been truly terrifying. While making the first-ever incursion up a river in Kalimantan in south Borneo, we'd pitched camp eight feet above the water. The flood came in the middle of the night while we were sleeping, rose ten feet before we knew it, and hammered right through the camp, carrying away all our possessions and leaving us shivering and terrified in the bushes above, without sleeping bags or tents to ward off the rain. Such close calls have left me with a more than healthy respect for the power of such forces of nature.

The steep walls of Mageni and its four-kilometre-diameter cauldron do not represent much of a water catchment area, but that geography meant that the river's reaction to torrential storms was close to instantaneous. Our tame river became the Styx within an hour, and was rising by the second.

As I pondered this, I was suddenly swept off my feet again, and lost my grip on the rope. I toppled into the water, scrabbled back up with one hand and caught hold of the line with just a couple of fingers. It all seemed so unreal. I was being dragged under, and in the back of my head a little voice was saying: 'That rotten bit of driftwood you secured the line with is never going to hold.'

This was it. I was going to die, I was going to get swept away into the maelstrom and never be seen again. I was going to be killed 'cos my rope techniques were too damn shoddy – Tim would be furious when he found out. It wasn't frightening, in fact I wanted to giggle at the fact the whole thing felt like something out of a straight-to-video B movie. Luke and Maxi were screaming and yelling just feet away from me, holding out their hands for me to grab them. Headtorches were cutting beams through the darkness, reflecting yellow in the stream and laser shows in the spray. I looked up at my hand grasping the rope – and then I managed to swing the other one up, grasp the line, and drag myself up to where Luke's and Maxi's hands were waiting.

'Fuck, bru, I thought you were a gonner,' said Luke. Maxi's eyes were even wider than when he thought Gandalf and the cave trolls were real and living in Auckland. If it'd been Bruce Willis, he'd probably have come up with some line about being a turd too tough to be flushed, but I was gibbering like a circus chimp, too far gone for any suggestion of machismo.

So now we had a situation. None of us had been bothering with raincoats here – no garment on earth would keep out a rainforest downpour, and while you're yomping through the forest they'd be too hot and sweaty – so we were just wearing T-shirts and almost enjoying being constantly wet. This works fine while you're on the go, but when you stop at night and it's still hammering down, and you're at a greater height above sea level than the very highest summits back home in Britain, it stops being fun very quickly indeed. Luke and Maxi were already slightly blue in the lips and shivering uncontrollably, and I could feel the insidious beginnings of hypothermia too. We had nothing to protect ourselves, no shelter, nothing.

Steve had already headed back for camp and might not notice our absence for a good while. It might be an hour before anything could be done, and by then God alone knows what sort of a state we'd all be in. It seemed the only option was for me to head back over on the line to get some help. I didn't bother to try and wade, as the water was way over chest height. Instead I just let my body go in the stream and pulled myself hand over hand. If a big log had come down, I would have been killed; if the driftwood anchor had snapped free – likewise, game over. I couldn't hear Luke's and Maxi's yells of support as the water was raging too hard and I was spending too much time with my head submerged. However, the rope held, nothing plummeted downstream into me, and I made the other side intact. I then struck out towards camp, a man on a mission to rescue his comrades in need.

I found camp already mobilised. Steve Greenwood was way ahead of the game, and had organised some waterproof dry bags with sleeping bags, emergency chocolate and dry clothes (where he'd been hiding that gold dust I determined to grill him about later). As the rescue team made to step into the breach, I forbade the smaller members of the team to join us, as the flow could quite easily have swept them away, even at the riverbank. The cameras rolled as we headed down the slippery slopes, serious intent etched into every face. Luke and Maxi needed us, and we wouldn't let them down.

When we reached the crossing point, we could see them huddled together for warmth on the other side. By now the rain was easing off, and looking at the waves I was sure they'd been more intimidating just minutes before. There was no time for dawdling, though. I set to, showing off my rope skills, rigging a throwline and then a makeshift pulley system to winch the packages across that would save my friends' lives. It took no longer than ten minutes to rig the system and get the dry bags across, round about the time that Jonny Keeling came down to join the party, having set up a second camera. He looked at the river, then at me flexing my biceps and clearly relishing the role of lifesaver, then at Steve, and just said: 'Is this all just for telly, Steve?'

There was disappointment as well as incredulity in his voice. His words – or more accurately, his tone – shook me abruptly out of my zone and I looked down at my feet. The water that when I'd started rigging the pulley system had been at my shins was now maybe two feet away from my toes. Somehow, while I was focused on my job, the water level had dropped as fast as it had risen. When I looked back out into the flow again, I saw not a river of certain death but a river that was little more than an everyday challenge.

Ten minutes later, and it had subsided to such a degree that Luke was willing to risk wading back through. We fixed a rope around his waist for extra safety, and all hauled in, just in case he lost control of the handline.

But then there was Maxi. Steve was still talking in hushed tones about how terrified he was of the water and how it might make him do something really crazy. Just then, it seemed his prophecy would come true. Maxi tried the handline, but then clearly decided he didn't trust it. Instead, he leapt from the side of the river on to a boulder in mid-river, ignoring the handline completely.

'*No, Maxi!*' we yelled from the bank.

We'd been lulled into a false sense of security. After all we'd been through, was our story going to have a tragic ending? Was Maxi going to die, right in front of us? He stopped, wobbling on his flimsy rock perch and seeming to contemplate for a second, as if making up his mind what his options were. Then he leapt into the stream.

'*Maxi! No!*'

He landed on another rock, that one partly submerged. Then he bounced effortlessly on to another, then another, then sprang right across what remained of the torrent. He might have been playing hopscotch. He bounded through the shallows, then up on to our bank where he grabbed me by the hand, grinning his big silly grin. For the third time that night, I felt very daft indeed.

The next morning was an all-fired-up panic to try and get the camp taken down and packed up in preparation for our evacuation. We were up at dawn, and saw above the helipad clear skies, not a cloud in sight. Surely we couldn't get that lucky? As the morning wore on, though, and we started ferrying boxes and bundles out to the helipad, the clouds began to roll in and spells of drizzle blighted our spirits. Nick had been wearing a haunted look for most of the preceding fortnight, and had obviously been counting the hours, minutes and seconds until he could escape.

'Don't do this, man,' he implored the skies and the weather gods. 'Don't you do this to me. I can't be here any longer, man, I have to get out.'

Pacing back and forth like a wolf in a cage, with his piercing blue eyes, his overgrown beard showing patches of white and his habit of scraping his uncut hair skywards at times of stress, Nick had started to look a little crazed. Worse, between the downpours two helicopter runs manage to get in, taking away a huge load of kit and several of the crew but leaving us behind, stewing in the knowledge that we'd be trapped here with few or no supplies and no shelter. Nick looked like he was going to start attacking people with frustration.

Late afternoon, and finally the last few chopper flights got through. Nick, Gordie and I made a dash for it, frantically clambering aboard like the last evacuees from Saigon. As the heli lifted off, I looked down into the riverbed I would surely never see again, towards the camp that was already invisible behind the trees and the undergrowth and would return to the forest's grasp within a matter of weeks. We were cruising up past the echidna ridgeline, the furthest point we had explored, within a matter of seconds. We gazed out into the crater, stretching in every direction for several kilometres, terrain that looked as if you could stroll right across it in an afternoon; but knowing what I knew now, I reckoned it would take a pretty dangerous and debilitating week.

All the wonders we had discovered had been found in a minuscule area of the caldera, and were doubtless just a tiny fraction of the miracles that lurked hidden in this Lost World. Maybe the Tasmanian tiger really was still living here. Maybe that new form of life that Wallace lusted after could be there too, in some forgotten corner of Bosavi. The flip side of this, though, was the knowledge of what man with his ingenuity and technology could do to this amazing natural resource. Heli-logging could strip it of timber in a few months. Tribal hunting parties could eradicate the wildlife in a year.

Bosavi was tangible proof that old-fashioned exploration was still possible in the generation of the God particle. There are still hidden corners of our planet that have defied all attempts to tame them, and some of them are truly wondrous. While Nick beside me sat grinning massively, elated to be escaping, I was honestly a bit torn. What other miracles and grand discoveries remained within its vertiginous prison walls? I didn't want anyone else to see them first! It remains a hidden gem of extraordinary beauty that provided a salve for my soul, and quelled the hurt of all my early failures. But Bosavi was also our world and our rainforests in microcosm, infinitely precious yet infinitely vulnerable. It had been the great privilege of my life to be there, to be the first Westerner into the volcano, but it would also be my curse, the standard to which every other adventure in my life would ever after be judged against.